Biolegalities

Series Editors

Marc de Leeuw
Law
University of New South Wales Sydney
Sydney, NSW, Australia

Sonja van Wichelen
Sociology and Social Policy
University of Sydney
Sydney, NSW, Australia

This interdisciplinary series on *Biolegalities* engages with contemporary challenges and implications of new biotechnologies and biological knowledges in the field of law. Our series aims to open up a much broader understanding of biolegality that includes a range of biotechnologies and biological knowledge, expanding into areas of immigration law, trade law, labor law, environmental law, patent law, family law, human rights law, and international law. While the growing scholarship on biopolitics has studied the ways in which such practices are entangled with certain modes of governance and neoliberal economies, their translations, deployments, and reconfigurations in the realm of law or legal practice has been relatively understudied. The main objective of this book series is to provide a venue for the study of the complex and often contested ways in which biotechnologies or biological knowledges are reworked by, with, and against legal knowledge.

More information about this series at
http://www.palgrave.com/gp/series/15629

Marc de Leeuw · Sonja van Wichelen
Editors

Personhood in the Age of Biolegality

Brave New Law

palgrave
macmillan

Editors
Marc de Leeuw
Law
University of New South Wales Sydney
Sydney, NSW, Australia

Sonja van Wichelen
Sociology and Social Policy
University of Sydney
Sydney, NSW, Australia

Biolegalities
ISBN 978-3-030-27850-2 ISBN 978-3-030-27848-9 (eBook)
https://doi.org/10.1007/978-3-030-27848-9

Cover credit: imagenavi/Getty Images

This Palgrave Macmillan imprint is published by the registered company Springer Nature Switzerland AG
The registered company address is: Gewerbestrasse 11, 6330 Cham, Switzerland

In memory of Sam, a dear colleague and bright scholar who left us much too early.

Acknowledgements

This edited collection is the result of an international workshop *Brave New Law! Legal Personhood in the New Biosciences*, which was held in late August 2018, at the University of Sydney and organized by the Biopolitics of Science Research Network. The workshop brought together 33 scholars from around the world to discuss changes in and transformations of legal personhood considering recent bioscientific developments. The premise of the workshop was that complex issues were emerging in the intersection of law and biology, an area of law that was quickly expanding and displaying highly composite problems that challenge existing legal regimes. Key questions included: What do the new biosciences do to our social, cultural, and legal conceptions of personhood? How does our legal apparatus incorporate new legitimations from the emerging biosciences into their knowledge systems? And what kind of ethical, socio-political, but also scientific consequences are attached to the establishment of such new legalities? By bringing together legal scholars, anthropologists, sociologists, historians, cultural theorists, and political philosophers, the aim of the workshop was to examine these problems by looking at materialities, the posthuman, and the relational in the (un)making of legalities, but also to critically assess the "newness" of these legalities, and to compare them with earlier investigations of natural personhood.

Fourteen papers were presented, and fourteen other international scholars gave generous comments that helped develop the papers presented in this volume. We would like to thank them here for their insightful papers, considerate reading of participants' papers, insightful feedback, and lively discussion: Courtney Addison, Sascha Callaghan, Margaret Davies, Nadine Ehlers, Jennifer Hamilton, Fleur Johns, Isabel Karpin, Hannah Landecker, Thomas Lemke, Cressida Limon, Neil MacLean, Maurizio Meloni, Catherine Mills, Karen O'Connell, Bronwyn Parry, Shobita Parthasarathy, Vincenzo Pavone, Brad Sherman, Margrit Shildrick, Halam Stevens, Cameron Stewart, Samuel Taylor-Alexander, Catherine Trundle, Britta van Beers, Robert Van Krieken, Miguel Vatter, Ayo Wahlberg, and Catherine Waldby. Our conversations continued at the Annual Meeting of the Society for the Social Study of Science which was held a few days later at the Convention Centre in Sydney and where the editors organized two consecutive panels on legal personhood and the new biosciences. More thought-provoking papers were presented in these panels, and we thank Seamus Barker, Shun-Ling Chen, Myra Cheng, Zsuzsanna Ihar, Jaya Keaney, and Declan Kuch for their participation. We are very grateful to Torsten Heinemann who took up the role of discussing these 4S papers. A special thanks to David Delaney—who was not able to come to the workshop—but who generously offered and provided a truly engaging afterword to our collection.

The workshop was part of the Biolegality Pop-Up Research Lab, an initiative from the Sydney Social Science, Humanities, and Arts Research Centre (SSSHARC) and Sonja would like to thank its director Professor Nick Enfield for making biolegality a research priority in the faculty and for generously funding the initiative. Claire Stevens and Zsuzsanna Ihar were indispensable in successfully preparing the pop-up events and were essential to making the workshop run smoothly and to lively (social media) engagements with people in and outside of the workshop. Sonja would also like to acknowledge the funding she received from the Australian Research Council for Early Career Researchers (DECRA, project number DE140100348) which allowed the initial ideas and writing of this project. She also acknowledges the SOAR scheme at the University of Sydney which generously allowed her teaching buy out to work on this project.

Marc would like to thank UNSW Law for supporting the *Initiative for Biolegalities*, of which he is the convener, and the financial support for several symposia between 2015 and 2018 (*The "Biological Turn" in Law: A Critical Appraisal*; *The Seeds of Law: Monsanto, Patents and the Public Good*; and *Natural and Unnatural Threats: Pandemics & the Global Governance of Health and Biosecurity*). He would also like to thank all participants whose talks and debates greatly enhanced his understanding of what is at stake in the biolegal dynamic. Marc also thanks the newly created UNSW Allen's Hub for Technology, Law and Innovation (where he is research leader for the research stream "Hybrid Life and Legal Personhood") for supporting his work.

In preparing the manuscript we would like to thank our editor Rachel Daniel for taking on this project, for her intellectual commitment to supporting scholarship from science and technology studies, and for her meticulous reading of the manuscript. Madison Allums has been extremely helpful in attending to our questions and last-minute requests. The research assistance from Pradytia Pertiwi has been exceptional. Sonja would like to thank her for her keen enthusiasm and meticulous work in bringing the manuscript to completion.

Contents

Notes on Contributors

Courtney Addison is an anthropologist of science and medicine, and Lecturer at the Centre for Science in Society at Victoria University of Wellington, New Zealand. Her research has explored gene therapy clinical trials through ethnographic research in a European hospital, and more recently, conservation genetics for the native species of Aotearoa New Zealand. She is interested in "everyday ethics" in scientific contexts, and how the genetic sciences are used to make claims to belonging and ownership.

Seamus Barker is a Ph.D. candidate in the Department of Sociology and Social Policy at The University of Sydney, Australia, where he is supervised by Professors John Frow and Danielle Celermajer. He is also supervised by Professor of Clinical Neuroscience G. Lorimer Moseley of the University of South Australia. Seamus previously worked as a physiotherapist in pain clinics, and his research investigates competitions between different paradigms of pain. He is interested in critical medical humanities, social theory, narrative theory, hermeneutics, and psychoanalysis.

Margaret Davies is Matthew Flinders Distinguished Professor of Law at Flinders University, Australia. She studied in Adelaide and the UK in the 1980s, and has written on a range of areas in general legal theory: notably property theory, legal pluralism, and feminist legal theory. She is the author of several books, most recently *Law Unlimited: Materialism and Pluralism in Legal Theory* (Routledge, 2017).

Marc de Leeuw is Senior Lecturer at the Law School of the University of New South Wales and convener of the UNSW Law *Initiative for Biolegalities.* His work engages with continental and analytical traditions in philosophy while focusing on questions of human agency, epistemological practices and ethics. His projects are often interdisciplinary and examine the intersection between the ethico-political and moral-legal fields. His book *Homo Capax. Paul Ricoeur's Renewal of Philosophical Anthropology* is forthcoming with Rowman & Littlefield.

David Delaney teaches in the department of Law, Jurisprudence and Social Thought at Amherst College, Massachusetts, United States. He is the author of *Race, Place and the Law* (University of Texas Press, 1998), *Law and Nature* (Cambridge University Press, 2003), *Territory: A Short Introduction* (Blackwell, 2005) and *The Spatial, the Legal and the Pragmatics of World-Making: Nomospheric Investigations* (Routledge/GlassHouse, 2010). He is also the co-editor of the *Legal Geographies Reader* (Blackwell, 2001), and of *The Expanding Spaces of Law* (Stanford University Press, 2014).

Nadine Ehlers teaches in the Department of Sociology and Social Policy at the University of Sydney, Australia. Her research broadly focuses on the socio-cultural study of the body, law, and biomedicine, to examine racial and gendered governance. She is the author of *Racial Imperatives: Discipline, Performativity, and Struggles Against Subjection* and *Deadly Biocultures: The Ethics of Life-making* (with Shiloh Krupar), and co-editor of *Subprime Health: Debt and Race in U.S. Medicine* (with Leslie Hinkson).

Zsuzsanna Dominika Ihar is a postgraduate researcher within the Department of Sociology and Social Policy, at the University of Sydney, Australia. She has previously worked on the uptake of affect and racialised mimicry by policing bodies in digital spaces; the delineation between citizen/non-citizen workers via protection from/vulnerability to zoonotic illnesses; as well as the racial histories behind contemporary techniques of bone densitometry. She is currently examining the discourses around environmental remediation in the Caspian region, as well as tracing alternate ontologies, mythologies, and archives of oil.

Bronwyn Parry is Professor of Global Health and Social Medicine and Director of the School of Global Affairs at Kings College London, the

United Kingdom. She has published extensively on assisted reproduction, dementia, bioinformation and biosurveillance, bodily commodification, posthumanism, legal approaches to the regulation of nature, and the public understanding of science. She was a member of the Nuffield Council on Bioethics from 2007 to 2013, is an editor of Economy and Society and an elected Fellow of the Royal Society of the Arts (FRSA).

Delphine Rabet holds a Ph.D. in International Political Economy from the University of Sydney and is currently finishing a Juris Doctor at the University of New South Wales, Australia, in order to practice law. Previously she studied a Master of International Relations and holds undergraduate degrees in Commerce and Economics. She has lectured in Politics at Macquarie University, UTS, and at The University of Sydney where she continues to work on an ARC research project investigating the political economy of corporate tax evasion.

Dean Southwood is currently finalizing a Juris Doctor at the University of New South Wales, while finalizing his Ph.D. in Bioinformatics and Genomics at the department of Molecular Sciences, Macquarie University, Sydney, Australia. He also holds a Bachelor of Advanced Science in Physics, Chemistry, Mathematics and Biomolecular Sciences and a Master of Research in Quantum Physics from Macquarie University. Dean has taught courses across a wide range of sciences at tertiary level, and has a keen interest in the interdisciplinary aspects of law and science.

Catherine Trundle is Senior Lecturer in cultural anthropology at Victoria University of Wellington, New Zealand. Her primary research interests include contested illness and environmental health, social inclusion and exclusion, inequality and access to health care, kinship, and ethics. Catherine's research on nuclear test veterans engages with anthropological ideas of contested illness, exposure, risk, responsibility, proof, in/visibility, and uncertainty. Her most recent book co-edited with Susanna Trnka, *Competing Responsibilities* (Duke University Press, 2017) critiques the politics of responsibilisation in diverse contexts.

Britta van Beers is Associate Professor at the Free University in Amsterdam, The Netherlands. A recurring theme in her work is the legal relationship between persons and their bodies in various biomedical contexts, such as assisted reproductive technologies, markets in human body materials and biomedical tourism. Recent publications include the co-edited volumes *Humanity in International Law and Biolaw* (CUP, 2014) *Symbolic Legislation and Developments in Biolaw* (Springer, 2016) and *Personalised Medicine, Individual Choice and the Common Good* (CUP, forthcoming in 2018).

Sonja van Wichelen is Associate Professor with the Department of Sociology and Social Policy at the University of Sydney and Director of the Biopolitics of Science Research Network. Her research broadly engages with the body, law, and science in the age of globalization. She is the author of *Legitimating Life: Adoption in the Age of Globalization and Biotechnology* (Rutgers University Press, 2018) and *Religion, Gender and Politics in Indonesia: Disputing the Muslim Body* (Routledge, 2010).

Miguel Vatter is Professor of political science with the College of Business, Government and Law at Flinders University, Adelaide, Australia since 2018. He previously taught at UNSW and has held teaching appointments in the US, Chile, Germany and China. His areas of research are republicanism, biopolitics, and political theology. His most recent book is *The Republic of the Living. Biopolitics and the Critique of Civil Society* (Fordham UP, 2014).

Abbreviations

AI	Artificial Intelligence
ALRC	Australian Law Reform Commission
ARTs	Assisted Reproductive Technologies
CBSA	Canada Border Services Agency
COPE	The Committee on Publication Ethics
CRISPR	Clustered Regularly Interspaced Short Palindromic Repeats
DBS	Deep Brain Stimulation
DNA	Deoxyribonucleic Acid
DTC-GT	Direct-to-Consumer Genetic Testing
EHRs	Electronic Health Records
eMERGE	Electronic Medical Records & Genomics Network
FDA	US Food and Drug Administration
GDPR	General Data Protection Regulation
GHRH	Growth Hormone-Releasing Hormone
GINA	Genetic Information Nondiscrimination Act
HIPAA	Health Insurance Portability and Accountability Act
HITECH	Health Information Technology for Economic and Clinical Health Act
HPPP	Human Provenance Pilot Project
IPR	International Property Rights law
IRBs	Institutional Review Boards
IVF	In Vitro Fertilization
IVG	In Vitro Gametogenesis
MoD	Ministry of Defence
MRI	Magnetic Resonance Imaging
NRPB	National Radiological Protection Board

ONT	Oxford Nanopore Technologies
PGD	Preimplantation Genetic Diagnosis
PTSD	Post-Traumatic Stress Disorder
RNAs	Ribonucleic Acids
SNPs	Single-Nucleotide Polymorphisms
STIs	Sexually Transmitted Infections
UK	United Kingdom
UKBA	United Kingdom Border Agency
U.S.	United States
WGS	Whole Genome Sequencing
WIRCA	Workplace Injury Rehabilitation Compensation Act 2013
WPI	Whole Person Impairment

CHAPTER 1

Brave New Law: Personhood in the Age of Biolegality

Marc de Leeuw and Sonja van Wichelen

In June 2019, Jeannie Suk Gersen wrote an article in *The New Yorker* entitled "How Fetal Personhood Emerged as the Next Stage of the Abortion Wars."[1] In her piece, she described the arguments put forward by Justice Clarence Thomas who commented on a recent abortion case. Gersen gave an account of how, in his striking appeal, Thomas likened abortion rights to a form of racist eugenics revivalism and invoked the authoritative framework of legal anti-discrimination norms. Rather than slipping into the polemical agenda set up by Thomas, Gersen points to its function as a distraction to the real issue at hand: the emergence of fetal personhood. Our volume focuses on such arrivals of personhood by looking through a *biolegal* lens. The "age of biolegality" in our title suggests that we have entered a

M. de Leeuw (✉)
Law, UNSW Australia, Sydney, NSW, Australia
e-mail: m.deleeuw@unsw.edu.au

S. van Wichelen
Sociology and Social Policy, University of Sydney, Sydney, NSW, Australia
e-mail: sonja.vanwichelen@sydney.edu.au

M. de Leeuw and S. van Wichelen (eds.),
Personhood in the Age of Biolegality, Biolegalities,
https://doi.org/10.1007/978-3-030-27848-9_1

phase in which biolegal formations increasingly dominate social and political changes. This situation triggers an urgent need to reflect on these changes and examine the sophisticated (re)negotiations between biology and law that inform new understandings of natural and legal personhood.

Biolegality is not a new term. The concept originates from the work of Michael Lynch and Ruth McNally (2009), where they describe the coproduction of biotechnology and legislation within the context of forensic science—a field that in the past few decades has been redefined by the introduction of genetic technologies. As Lynch and McNally show, law, and law enforcement, play a unique role in communicating the power of DNA in situations of claims-making and crime. The exceptional reliance on DNA as a "truth machine"—they argue—indicate that biolegality is not only a "historical relationship between biological innovation and enabling legislation, [rather,] it is an epistemic relation in which biological 'truth' justifies exceptional legal procedures" (2009, p. 296). Here, biolegality is set against the context of a more general process of geneticization in society (Hedgecoe 1999; Lippman 1991; Nelkin and Lindee 1995), where following the human genome project, genetics succeeded to consume the everyday, and informed basic knowledge about health, disease, kinship, and identity. Lynch and McNally's biolegality hinged on this primacy of genetics and the focus on knowing.

In this collection, we stress that the current age of biolegality extends the field in two ways. First, we have entered a postgenomic phase, where genetic reductionism is challenged both in and outside of the life sciences through the biological conceptualizations of a more permeable gene interaction (Richardson and Stevens 2015; Meloni 2016, 2019). Without denying that genetics still plays a dominant role in biological thinking, its operation is fundamentally complicated by discoveries that their phenotypical expression is highly dependent on a developmental system, rather than genetic structure alone (Griffiths and Stotz 2006). Within this new epistemic framework, the environment is vital to understanding the operation of genes. Epigenetics, for instance, is now increasingly introduced as a key variable to diagnostics in biomedicine on the one hand (for instance in personalized medicine) and incorporated into stories of kinship and identity on the other (see Warin et al. 2018).

Second, while *knowing* life and the value of knowledge itself stood at the center of the genetic project, it is the *making* and *engineering* of life that more and more defines biolegality in the present time. Generative medicine and gene editing are just two of the examples taken up by some of our authors. Here, law is confronted not only with the way biology is being used or manipulated but rather with the way it is fundamentally remade. As the chapters in this book illustrate, the generative potential of this shift from *knowing* to *making* impacts our legal tools and reconfigures what we mean by the person.

Law does not just regulate, allow, or limit what can be done or undone biologically, but also defines or alters our philosophical, political, or social self-understanding (Delaney 2003). In law persons are functional fictions with a real effect—they can be natural or artificial (and to complicate matters just a bit more, they can be artificially natural). As "practices of knowledge," both law and biology define or construct persons according to a specific idea of "human nature," "bodies," "organism," "groups," or "life" (to name a few). This construction of persons—so David Delaney explains in his afterword to our book—often implies an undoing or redoing of established concepts of personhood. Through such a multifaceted understanding of persons, this collection is concerned with how biology (and nature) determine what law and legality can *be* and *do*, and vice versa, how law and legality determine what biology can *be* and *do*.

Personhood is enacted in many ways; this volume investigates the sociopolitical, symbolic and material circumstances that allow (or limit) the idea of persons, personality, or personhood, taking into account the discursive and performative ways in which persons are fashioned. While some contributors take legal events as a starting point to leverage questions about biological citizenship (Trundle), racial identity (Ehlers), or molecular governance (Ihar), others take the thought-provoking task to go deeper into legal theory (Van Beers, Vatter, Davies) and to address not only the legal circumstances through which personhood is mounted, but also to propose alternative frameworks that place personhood in a better shape for some of the future (biological) challenges awaiting us.

In this introductory chapter, we provide a background to the conceptual framework of biolegality and chart out the structure of the book by discussing the scholarship informing our inquiries around personhood today. The tenet of the book is not—at least not in its first principle—to articulate normative ways of approaching personhood in the age of biology. Instead, the contributions reveal the messiness and complexities underpinning the active shaping of persons—whether this pertains to the apparatus of biomedicine, and the ways they are entangled with legal and governmental practice, or to how individuals and communities invoke old and new tropes of persons amidst their claims for rights, recognition, and citizenship.

Legality in the Age of Biology

Despite the risk of being perceived as fetishizing the bio in our approach to legality, and thereby joining the growing group of scholars invested in mapping biopolitics, biosociality, biocitizenship, biolegitimacy, biocapital, and bioeconomies, our edited collection focuses and advocates the concept of "biolegality" to give an account of multiple reworkings of law's persons within different formations: law *as* biology, law *and* biology, law *in* biology, law *through* biology. We define biolegality as a knowledge practice where law or legality—in its foundational negotiation with biological form, practice, or reason—allows the stabilization of epistemological and ontological ideas of biology, nature, life, materiality, and sociality. Against the dichotomy between a rigorous legal pluralism, and a strict conception of sovereign law or legal governance, our conception of biolegality recognizes the social construction of law on the one hand, but at the same time, acknowledges that the law—with its legal forms, doctrines, and techniques—has a distinctive quality. Similarly, biology too inhabits unique forms of biological reason and practice, which define their understandings and applications in society today.

In her contribution to our volume, Margaret Davies aptly describes this productive tension inherent to biolegality, which is "neither the application of law to bioscience nor an account of the fundamental laws of biology, but rather the generative and regenerative formations of both law and life: the patterns, iterations, symbiotic relations, responses, and behaviors of life understood as law" (Chapter 12, this volume). We extend this understanding beyond the formal confines of law, and into the broader field of legality, which includes practice and discourse outside of courts and institutions, and incorporates meaning-making practices and authoritative sources that are "recognized as legal, regardless of who employs them or for what ends" (Seron and Silbey 2004, p. 51; Silbey and Ewick 1998). Rather than a social institution, legality is a heuristic tool, an analytical term that is useful to locate the enactment of law outside of its official structures.

Concurrently, technoscientific developments in the realm of biology are fundamentally testing the templates of legal formulas and techniques (Pottage 2007). The emergence of fetal personhood with which we started this introduction is an apt example where medical technologies such as obstetric ultrasound but also MRIs and diagnostic testing are increasingly implicated in the ethical and legal evaluation of fetal or reproductive rights (see Dumit 2004; Timmermans and Berg 2003; Mills 2014). The point that we want to make with biolegality, however, is that the technologies involved do not merely, or simply, add another dimension to existing legal conventions around personhood, rather, biotechnology fundamentally tests the *premises* on which these legal notions were created and constituted. The science and technology studies scholar Sheila Jasanoff writes in this respect about "bioconstitution" (2004) to argue for the co-constitutiveness of science and society. Law represents a core element in the *conditioning* of science in society. Since regulations form an integral part of this process, the work of law and that of its actors (lawyers, judges, and lawmakers) are tantamount to capture the full breadth of the translation of science into society. It is only in such a way that we can decipher the different languages, methods, and interpretations that scientists and lawyers use to give meaning to developments in science. These approaches surpass the actual institutions of science and law as they exist in our everyday lives and culture.

While some of the contributors discuss how genes, chromosomes, phenotype, or brain plasticity challenge our assumptions of personhood, others contemplate how new knowledges stemming from the biosciences are put to work in legal or regulatory contexts that "make up people," and *create* novel and certain *kinds* of persons (Hacking 2006, 2007). Other chapters have focused on emerging findings in postgenomic science, in the field of epigenetics, the human microbiome, synthetic biology, as well as technologies such as gene editing that anticipate adjacent problems in demarcations of the bodily self and organic life. Furthermore, as Southwood indicates in his contribution, today's direction in the medical world toward "precision" or "personalized" medicine, revisits the legal understanding of the personal and the way rights and responsibilities are given due consideration (see also Van Beers et al. 2018).

Finally, most of our authors stress a posthuman incentive to look anew at the work that personhood does in the imagination of our social and legal institutions. In the context of the Anthropocene, the question of the posthuman goes beyond the focus on anti-humanism and is posed within the reconfiguration of life itself (Braverman 2015; Grear 2010; see also De Leeuw and Van Wichelen, forthcoming). Sustained collaboration between biolegality and science studies allows us to theorize the relations across biology and law, which proves useful to examine the populations and bodies in their entanglements with the ecological, geological, and environmental.

The Subject of Personhood

The subject matter of personhood—and more specifically legal persons—has received renewed attention in the past decade (Van Beers 2017; Grear 2013; Mussawir and Parsley 2017; Naffine 2009; Pottage 2002, 2004, 2007; Vatter and De Leeuw 2019; Wishart 2016). Ever since Roman law, the project of personhood is that of a pure legal construction. Within law, the meaning of personhood did not derive from any idea or knowledge about biological or social individuality. Instead, "differentiated transactional personae" were the commonplace of legal personhood in Roman times, denoting how fictions were used to execute economic exchange including arrangements concerning inheritance and wealth (Pottage 2004, p. 30). It was not until the influence of Christianity that the fictional person fused with a biological and unitary concept of the person (ibid.; see also Naffine 2009).

What evolved into what we now know as legal personhood in Western law can be understood as a process of artificialization on the one hand and naturalization on the other (Van Beers 2017). Natural personhood came to be increasingly imbued with bioscientific understandings of individuality, constructing the human and the self from the science of knowing the human body. Existing legal tools that mark off the boundaries of the body are weakening considering biomedical advances (Jasanoff 2016; Hoeyer and Hogle 2014; Radin 2018). Our contributors attest to the fragmentation of natural and legal personhood and discuss what it means for society at large. While the coping of this fragmentation does not always take place within the proper institutional structures of the law, but in the contexts of everyday social or medical practices, biomedical technologies and expertise increasingly enter the courts and account for compound jurisprudence.

For instance, the case of fetal personhood demonstrates how technologies such as obstetric ultrasound and brain imaging exercise power in "picturing" personhood; while for the layperson biomedical scans and images conjure objective truth, anthropological scholarship indicates that the reading or consumption of such imaging technologies are really a complex cultural and technoscientific process (Dumit 2004; see also Mills 2014). These processes filter into legal practice and are muddled further by the fictionalizing qualities of legal persons.

Such fictionalizing qualities of natural personhood can be stretched even further in contemporary cases of "wrongful birth." Here, biomedical advances in prenatal diagnosis (including preconceptive diagnosis) have led to legal situations where "future" or "intended" children are deliberated, stretching the person to the point of preconception to "future persons" (see our contribution in this book and also Van Beers 2017). These "stretchings" or further *artificialization* of personhood are not always innocent. Some of our contributions attest to the biopolitical dynamics inherent to reconfiguring personhood in courts or forensic practice (Ehlers and Ihar). Others focus on the limits of persons, pointing critically at the bounded idea of the individual and bodily autonomy foundational to legal concepts of personhood. As Addison in this volume argues, autonomy is continually compromised by people's social commitments, bodily frailties, and the power dynamics of medicine. While the physical person may be anchored in a locale and a certain time period, their endurance through virtual and electronic technologies point toward a continuity of personhood beyond the personal body.

The different dimensions and methods to the analysis of personhood in the age of biolegality have prompted four delineations of persons: *troubling persons* (Chapters 2–4), *evidencing persons* (Chapters 5–7), *governing persons* (Chapters 8–10), and *the future of persons* (Chapters 11–13). Each part consists of contributions that describe specific boundaries or new allowances afforded to the notion of personhood, ranging from the bodily, to the virtual, the molecular, the chemical, the prosthetic, the plastic, and the epigenetic. While some chapters can easily be categorized in more than one group, cutting across the modes of troubling, evidencing, governing, or futures of persons, the parts situate the different analytics and stakes involved in examining personhood and allow for a more careful reading of their configurations vis-a-vis bioscientific and biotechnological developments.

Troubling Persons

The first part of the book attends to how developments in the biosciences or biotechnology are troubling, fragmenting, or rearranging legal conceptions of personhood. These developments fundamentally disturb the core principles of natural personhood in law (like liability or culpability, responsibility, rationality, choice, intentionality, and identity). Parry's contribution, for instance, examines three pertinent areas in contemporary biomedical advances to highlight these rearrangements: bioinformation, regenerative devices, and oocyte donation. She tracks how personality came to exist in virtual registers, thereby disrupting historical understandings of the generation and transmission of personality. The conceptual frameworks of dominion, culpability, and belonging are central to these reworkings, and Parry proposes that we sustain these values in the current situation where personalities increasingly lose a clear genealogy or legal standing.

Expansions in the biosciences do not evolve in a political or social vacuum and cutting across the analyses of the contributions are political conditions that make the social environments conducive to leveraging biotech. As such, neoliberal discourse and ideology are at the heart of the economies driving the bioscientific and biomedical world today. Here, life has become properly appropriated by capitalist processes of accumulation (Cooper 2011; see also Rajan 2006). While the emergence of *bioeconomies* concerned pharmaceutical or biotech companies in the first instance, more intricate markets in the field of reproduction (assisted reproductive technologies, surrogacy, and egg banking) or genetic testing also moved into this neoliberal space (see Pavone and Goven 2017).[2] In true Foucauldian fashion, contributions in this section demonstrate how neoliberal structures and agendas inform the notion of personality and the self at the core of natural personhood. Such representations of the self—whether virtual or plastic—fit all too well with neoliberal understandings, including the maximizing, entrepreneurial, and responsible legal subject. Within a neoliberal framework, individuals are seen—and see themselves—as subjects responsible for their health and health optimization.

The latter is especially evident in Rabet's contribution where she describes how developments in neuroscience contribute to new ways of thinking about human behavior, specifically with regards to notions of responsibility and accountability. Through the analytical lens of political economy, Rabet traces how developments in neurolaw fit comfortably with the contemporary neoliberal ethos. While neuroscience can also offer sites of contestation that resist some of the neoliberal narratives, their effects are still quite minimal. Rabet concludes that contemporary co-productions of neuroscience and law, which displace responsibility, hinders the legal person by ignoring the interaction of biological, sociological, and psychological factors in the explanation of human behavior.

Finally, running through several contributions in this volume, are posthuman analytics framing the current reworkings of personhood in law and biology. The posthuman—as incarnated through Foucault's anti-humanist position—has been key to informing biopolitical analyses of neoliberalism, which critiqued the reliance on liberal reason and rationalities of Enlightenment to further the universal political subject. The posthumanist turn that legal scholars have recently taken, however, goes beyond the focus on anti-humanism and is posed within the reconfiguration of life itself: how can life be governed when the boundaries of life are shifting, when inanimate, nonhuman, and posthuman forms and their novel ontologies have entered the specter of qualified life, which was mostly attributed to humans in the classic understanding of biopolitics. This question is heightened in the age of the Anthropocene, where the destruction of the natural world is evidenced by the burden humans have put on natural resources, land, soil, waters air and nonhuman life, including plants and microbes. It is under these new conditions that questions around legal and natural personhood are newly posed and analyzed.

Our own contribution to this book examines this posthuman troubling of personhood by way of engaging with biological knowledge stemming from postgenomic findings. Such findings in the area of epigenetics, research on the human microbiome and immunology are challenging the perception that bodies, brains, genes, and cells are bounded entities. Instead, postgenomic knowledge defines biological life as plastic. Here, biological processes are more permeable and malleable, rather than linear and determinant. In our chapter, we compare these approaches to biological plasticity, with approaches to plasticity in the study of law's persons via the concept of "entityhood." We argue that the materialization of legal entityhood remains fraught since it does not allow room for posthuman relationality. Our analysis suggests that plasticity—both in law as science—runs the risk of fragmenting personhood to the point of incapacity and continues to reiterate Euro-American and modern assumptions of property.

Evidencing Persons

This section focuses on local, national, and transnational cases that demonstrate how personhood becomes entangled in the pursuit of citizenship. Notions of community and identity are central to these pursuits, as are the concepts of rights and justice. Here, evidence and proof become attached either to the body directly, or to the biological information of the body, contributing to a form of "biological citizenship" (Rose 2007; Petryna 2013). It is essential to place the various new articulations of biological citizenship, and the involvement of DNA testing and medical screenings in and outside of the courts, against the backdrop of an ever-increasing biologization and geneticization of law and society more generally (Nelkin and Lindee 1995). Within these biolegal configurations, law becomes the means to arrive at "the truth [of] the body" (Fassin and d'Halluin 2005). While law continues to struggle with the incorporation of genetics and the constitutive force of genetic knowledge (Strathern 2005), other *molecular* forms of biological knowledge—such as epigenetic knowledge—increasingly enter the legal domain. This situation further fragments the structures of legal and natural personhood, and contributes to new forms of biolegitimacy (Fassin 2009, p. 52).

This is particularly true for Trundle's chapter. Drawing on research in New Zealand and the UK with military veterans of British nuclear tests, her chapter tracks test veterans into various social arenas to explore the politics of proof-making. Trundle argues that the task of making various forms of proof count, medically, legally, and politically, depends on how successfully different types of proof can translate across fields of knowledge. While the untranslatability of multiple modes of proof across these domains bolsters the state's hold over claims of truth, it also reinforces the opposition to it.

Trundle and Barker stress the role of narrative in the coproduction of law and science. But in contrast to the narratives shaped by state actors or test veterans in Trundle's case, Barker focuses on paradigms of pain and seeks to examine how nonhuman actants—such as "injury"—can be bound up in the same causal-moral economy as human characters' narratives of pain. By considering the moral dimension of narrative, his chapter examines how a paradigm—as a form of potential emplotment—can serve as a powerful force in the legal negotiations of personal injury. Moreover, Barker argues that these co-productions clarify the persistence of the body–mind dualism, despite non-dualistic paradigms in science and medicine.

Indirectly, proof and evidence figure prominently in Ehlers' discussion of reproductive technologies in case law. Drawing from the 2014 *Cramblett v. Midwest Sperm Bank*, Ehlers asks how someone's race—in this case a black child—can be utilized and evidenced as "injury" against a white parent. Rather than a form of agency, the biological evidence performed in court functioned as a claim of grievance. In her assessment of the biolegal arrangements, Ehlers explains how the judge could not proceed under a "wrongful birth" claim. While the ruling is seen as legitimating the value for black life, Ehlers instead argues that the court failed to address race at all and thereby renounces legal responsibility to address persistent forms of racism.

Governing Persons

This section draws on the continuing importance of (non)governance and (non)regulation. Ihar's chapter clearly stipulates the persistent role of the state and legal governance in determining which biological criteria are adequate to convey citizenship rights within the boundaries of the sovereign nation-state. Drawing further on existing scholarship on border security and immigration regimes that have attended to the increasing use of biometric and genetic technologies in managing the flow of people through national borders (Pugliese 2010; Heinemann and Lemke 2014; Heinemann et al. 2016; Van Wichelen 2017), her contribution highlights the uses of epigenetics to advance border security, and shows that rather than helping to extend rights it is restricting rights of citizenship. Focusing on the Hildesheim case, which involved the utilization of an epigenetic clock to verify age, and the incorporation of mitochondrial DNA and Y chromosome testing by the UK Border Agency to verify migrant claims of nationality, she examines how environmentally responsive genes turn into calculable, quantifiable, and ultimately depoliticized entities, and demonstrates how epigenetic language becomes methodologically useful for governance and administrative processing.

Addison's contribution appears to be the opposite of Ihar's chapter when comparing them on the grounds of governance. As a classic case of self-governance in the way of Foucault's *care of the self*, Addison describes how self-experimenters with posthuman and anti-aging aspirations have begun using CRISPR techniques on themselves, either through traditional experimental science or by live-streaming their actions. According to Addison, personhood—and the person–society relationship—are central to understanding the motivations behind these ventures. By historicizing self-experimentation practices, she examines how they have changed and shaped the governance of personhood. She argues that the limits of the person are reorganized by the coming together of genetic and digital tools that are at once biological (how long can we live?), social (who can we reach?), and regulatory (what can we do to ourselves and to others?). Such questions pose a challenge for the law, and how it should address the power relations of self-experimentation.

Within regular scientific labs, such experimentation would either be illegal or require long and arduous ethics approval. But by turning to their own bodies, Addison's biohackers apply the *care of the self* to its extremes; as they become the testing ground for scientific experiments, they also perform *biological* agency. Like hacking into your own computer, the hacking of your own body escapes regulatory scrutiny. Whether this practice should be viewed as courageous citizen-science, self-governing care, posthuman enhancement, or immature self-exploration is kept open. But what does get problematized are the boundaries and limits of personal exploration. In a sense, Addison's biohackers are working on the liminal boundaries of the body's status in law.

Southwood's chapter similarly picks up on regulatory challenges of personhood but focuses on the emerging field of precision medicine. The collection and storage of personal genetic data are central here and define some of the main concerns surrounding precision medicine. Becoming more valuable and therefore prone to exploitation, the privacy and security of personal data are at stake (see also Prainsack 2017; Van Beers et al. 2018). This raises the pertinent question of who the person in personalized medicine is? The plight of personal medicine—namely to use population data to tailor medicine to the individual—assumes control is given to the patient/consumer. However, as Southwood shows, this data—accrued from information put in health records, stored in social media (such as Facebook and Twitter), and digital applications (such as fitness apps, diet apps, or mental health apps)—is shared among multiple platforms and networks with the individual losing all control over one's own data. In his chapter, Southwood describes the associated problematic of the "right to know," where genetic information should be disclosed to third parties, that, in turn, can be accompanied by social, psychological or legal repercussions. While reforms are yet to take place in many jurisdictions, data sharing continues to increase across jurisdictions and in transnational corporate arrangements. Southwood presses for a more thorough discussion between the legal, scientific, and public community to diagnose key issues in precision medicine and to collaboratively find ethical answers to them.

The Future of Persons

While the preceding chapters looked to diagnose contemporary situations of modern personhood amidst developments in biotechnology, the chapters in the last section offer new ways of imagining future persons in law by way of taking these diagnoses to speak to legal theory. The chapters by Van Beers, Davies, and Vatter show different "registers" of the personhood/law nexus and give an account of the *genealogical-analytical* development of personhood more generally. Honoring the subtitle of our collection, they offer fresh perspectives to a Brave New Law capable of dealing with some of the pressing issues described in the former sections.

In her chapter, Van Beers looks at Gunther Anders' philosophical anthropology and his concept of "the obsolescence of human beings," which she contends are highly relevant for current debates about legal personhood. Retorting to some of the trends diagnosed in the preceding sections, Van Beers reflects on the growing tendency to further *artificialise* personhood in response to biotechnological developments. While *artificialisation* is regarded as the "liberation" of the legal person from its naturalistic confines, the gap between *homo* and *persona*—Van Beers argues—is widening. Working against the thought that the natural person is becoming obsolete, Van Beers employs Anders' framework to argue for the importance and continuance of the natural person in future jurisprudence.

The antecedent to the legal debate on personhood is the subject of property. And even though the preceding chapters only tangentially broached the subject, the topic is a vital component to the crisis of subjectivity and the ensuing ambiguity around the naturalization and artificialization of persons. As some of the chapters will show (particularly Rabet's and our contribution) contemporary late and global capitalist contexts collapse questions on personhood to the subject of property. As opposed to property and things (that can be owned), persons have rights, and these are protected by national and transnational jurisdictions. Ownership has always been key in reifying things and personifying persons, but the new conditions set out by biocapital and global bioeconomies are rearranging the premises around ownership and bodily material. In her contribution, Davies describes how normative liberal property theories—that assume a pre-social individual and its differentiation from a world of objects—are foundationally disturbed in a posthuman world. Moreover, echoing other contributions, she states that they are at the center of global ecocide and ongoing colonial practices. It is time, therefore, to reimagine property in an entangled world and Davies' contribution lays out some of the groundwork for a reconsideration of property by examining the abstract account of law and the profile of property that presently exist.

Vatter too makes a radical intervention in the reconfiguration of personhood in law. Similar to Van Beers and our own contribution, he observes that the turn to extending artificial personhood (for instance by entityhood) does not do justice to the ecological and posthuman aspirations in reconfiguring legal personhood. Though rather than strengthening or reemphasizing the importance of natural personhood (as Van Beers does), he suggests we look elsewhere. He proposes to bring back a distinction between corporation and trust, and to approach trust as a concept that is far more effective as a vehicle for the ascription of legal personality to nonhuman things than the ideas of fictional legal personality and corporate personhood. As Vatter demonstrates, the conception of trusts connects with developments in the bioeconomy as well as the Anthropocene and offers a radically different relation between use and ownership. By allowing "some third person to use what is owned by me; what is owned by me cannot be consumed by me because it places me under the obligation to sustain it" (see Chapter 13). Employing such an idea of trust will enable a radical new approach to "nature's law" and "law's law."

Finally, David Delaney truly honors the genre of the Afterword, by reflecting theoretically and intellectually on the chapters, and on the collection as a whole. He extracts the following core dilemmas of the book:

> A central question posed by the volume is: how might what we call law respond to this disintegration? Or, put another way: how does

> metaphysical dissensus translate into politics and how does the politics of personhood revision get re-translated into the idioms of rules and rights? How can law even maintain its always deferred dream of coherence when one of its foundational underpinnings is, like a broken mirror, crazed? (see Chapter 14)

The answers are a work in-progress. And rather than escape into an apocalyptic morass, Delaney points to the more tempered ways in which this volume envisions better persons.

Notes

1. See https://www.newyorker.com/news/our-columnists/how-fetal-personhood-emerged-as-the-next-stage-of-the-abortion-wars?utm_campaign=aud-dev&utm_source=nl&utm_brand=tny&utm_mailing=TNY_Daily_060519&utm_medium=email&bxid=5c7dc86c40f8667df249e39a&cndid=56616948&esrc=&utm_term=TNY_Daily.
2. We mean here intricate in that often implicate gift economies that are deemed noncommercial.

References

Braverman, I (ed.) 2015, *Animals, biopolitics, law: lively legalities*, Routledge, Abingdon.

Cooper, ME 2011, *Life as surplus: biotechnology and capitalism in the neoliberal era*, University of Washington Press, Seattle.

Delaney, D 2003, *Law and nature*, Cambridge University Press, Cambridge.

De Leeuw, M & Van Wichelen, S, forthcoming, *Biolegalities: A critical intervention*, Palgrave Macmillan.

Dumit, J 2004, *Picturing personhood: brain scans and biomedical identity*, Princeton University Press, Princeton.

Ewick, P & Silbey, SS 1998, *The common place of law: stories from everyday life*, University of Chicago Press, Chicago.

Fassin, D 2009, 'Another politics of life is possible', *Theory, Culture & Society*, vol. 26, no. 5, pp. 44–60.

Fassin, D & d'Halluin, E 2005, 'The truth from the body: Medical certificates as ultimate evidence for asylum seekers', *American Anthropologist*, vol. 107, no. 4, pp. 597–608.

Grear, A 2010, *Redirecting human rights: Facing the challenge of corporate legal humanity*, Springer.

Grear, A 2013, 'Law's entities: complexity, plasticity and justice', *Jurisprudence*, vol. 4, no. 1, pp. 76–101.

Griffiths, PE & Stotz K 2006, 'Genes in the postgenomic era', *Theoretical Medicine and Bioethics*, vol. 27, no. 6, pp. 499–521.

Hacking, I 2006, 'Making up people', *London Review of Books*, vol. 28, no. 16, pp. 23–26.

Hacking, I 2007, 'Kinds of people: moving targets', *Proceedings of the British Academy*, vol. 151, pp. 285–318.

Hedgecoe, AM 1999, 'Reconstructing geneticization: a research manifesto', *Health Law Journal*, vol. 7, pp. 5–18.

Heinemann, T, Helén, I, Lemke, T, Naue, U & Weiss, M 2016, *Suspect families: DNA analysis, family reunification and immigration policies*, Routledge, London.

Heinemann, T & Lemke, T 2014, 'Biological citizenship reconsidered: the use of DNA analysis by immigration authorities in germany', *Science, Technology, & Human Values*, vol. 39, no. 4, pp. 488–510.

Hoeyer, K & Hogle, LF 2014, 'Informed consent: the politics of intent and practice in medical research ethics', *Annual Review of Anthropology*, vol. 43, pp. 347–362.

Jasanoff, S (ed.) 2004, *States of knowledge: the co-production of science and the social order*. Routledge.

Jasanoff, S 2016, *The ethics of invention: technology and the human future*, W. W Norton, New York.

Lippman, A 1991, 'Prenatal genetic testing and screening: constructing needs and reinforcing inequities', *American Journal of Law and Medicine*, vol. 17, pp. 15–50.

Lynch, M & McNally, R 2009, 'Forensic DNA databases and biolegality', in *Handbook of genetics and society*, pp. 283–301.

Meloni, M 2016, *Political biology: science and social values in human heredity from eugenics to epigenetics*. Palgrave Macmillan, Basingstoke.

Meloni, M 2019, *Impressionable biologies: from the archaeology of plasticity to the sociology of epigenetics*, Routledge, London.

Mills, C 2014, 'Making fetal persons: fetal homicide, ultrasound, and the normative significance of birth', *Philosophia*, vol. 4, no. 1, pp. 88–107.

Mussawir, E & Parsley, C 2017, 'The law of persons today: at the margins of jurisprudence', *Law and Humanities*, vol. 11, no. 1, pp. 44–63.

Naffine, N 2009, *Law's meaning of life: philosophy, religion, Darwin and the legal person*, Bloomsbury, London.

Nelkin, D & Lindee, MS 1995, 'The media-ted gene. Stories of gender and race', in *Deviant bodies: Critical perspectives on difference in science and popular culture*, pp. 387–402.

Pavone, V & Goven, J (eds) 2017, *Bioeconomies*, Palgrave Macmillan.

Petryna, A 2013, *Life exposed: biological citizens after Chernobyl*, Princeton University Press, Princeton.

Pottage, A 2002, 'Unitas personae: on legal and biological self-narration', *Law & Literature*, vol. 14, no. 2, pp. 275–308.

Pottage, A 2004, 'Introduction', in A Pottage & M Mundy (eds), *Law, anthropology and the constitution of the social: making persons and things*, Cambridge University Press, Cambridge.

Pottage, A 2007, 'The socio-legal implications of the new biotechnologies', *Annual Review of Law and Social Science*, vol. 3, pp. 321–344.

Prainsack, B 2017, *Personalized medicine: empowered patients in the 21st century?* New York University Press, New York.

Pugliese, J 2010, *Biometrics: bodies, technologies, biopolitics*, Routledge, New York.

Radin, J 2018, 'Ethics in human biology: a historical perspective on present challenges', *Annual Review of Anthropology*, vol. 47, pp. 263–278.

Rajan, KS 2006, *Biocapital: the constitution of postgenomic life*, Duke University Press, Durham.

Richardson, SS & Stevens, H (eds) 2015, *Postgenomics: Perspectives on biology after the genome*, Duke University Press.

Rose, N 2007, *The politics of life itself: biomedicine, power, and subjectivity in the twenty-first century*, Princeton University Press, Princeton.

Seron, C & Silbey, SS 2004, 'Profession, science, and culture: an emergent canon of law and society research', in *The Blackwell companion to law and society*, Blackwell, Malden, pp. 30–59.

Strathern, M 2005, *Kinship, law and the unexpected: relatives are always a surprise*, Cambridge University Press, Cambridge.

Timmermans, S & Berg, M 2003, 'The practice of medical technology', *Sociology of Health & Illness*, vol. 25, no. 3, pp. 97–114.

Van Beers, B 2017, 'The changing nature of law's natural person: the impact of emerging technologies on the legal concept of the person', *German Law Journal*, vol. 18, no. 3, pp. 559–593.

Van Beers, B, Sterckx, S & Dickenson, D (eds) 2018, *Personalised medicine, individual choice and the common good*, Cambridge University Press, Cambridge.

Van Wichelen, S 2017, 'Reproducing the border: Kinship legalities in the bioeconomy', in V Pavone & J Goven (eds), *Bioeconomies: Life, technology, and capital in the 21st century*, Palgrave Macmillan, pp. 207–225.

Vatter, M & De Leeuw, M 2019, 'Human rights, legal personhood and the impersonality of embodied life', *Law, Culture and the Humanities*, https://doi.org/10.1177/1743872119857068.

Warin, M, Kowal, E & Meloni, M 2018, 'Indigenous knowledge in a postgenomic landscape: the politics of epigenetic hope and reparation in Australia', *Science, Technology, & Human Values*, https://doi.org/10.1177%2F0162243919831077.

Wishart, D 2016, 'Puzzling out law's person', *Indiana Journal of Global Legal Studies*, vol. 23, no. 2, pp. 743–764.

PART I

Troubling Persons

CHAPTER 2

Spectral Personas: Exploring the Constitution and Legal Standing of "Virtual Personhood"

Bronwyn Parry

Introduction: Concepts and Constructs

In this chapter my intention is to begin to unpack the complex relationship between personality and legal personhood in the brave new "post-human" world we now inhabit. As one lacking in legal qualifications I cannot undertake the task of suggesting the refinements in existing law that might be needed to accommodate an expanded conception of personality but rather seek here to establish how and in what ways the law might be called upon to take account of what I refer to as "spectral personas." To understand how existing concepts of personhood are challenged by developments in the biotechnological realm it is perhaps useful to begin by outlining how personhood is currently used, in law, as a device to construct entities to which particular rights and obligations are due. It would be perfectly understandable for the uninitiated reader to presume that the condition of legal personhood, and personality,

B. Parry (✉)
Kings College London, London, UK
e-mail: bronwyn.parry@kcl.ac.uk

M. de Leeuw and S. van Wichelen (eds.),
Personhood in the Age of Biolegality, Biolegalities,
https://doi.org/10.1007/978-3-030-27848-9_2

should always and only (as a normative precondition) be assigned to individual human beings. This is the conception that is captured in the legal construct of the "Natural Person," defined in law as an "actual human being" rather than an "artificial person." The extension of the rights of personhood to other entities (either human or nonhuman) that could be recognized as holding the privileges, rights and responsibilities normally accorded to a human being (for example rights of protection) was a logical and necessary next step even though it seemed surprising to some that a "Legal Person" could take the form of a collective of human beings, a nongovernmental organization or agency, a trust, partnership or municipality rather than an individual.

As the legal scholar Christopher Stone noted in his seminal 1972 paper; suggestions that rights of personhood be extended to non-humans was, for many, and initially, completely inconceivable. As he notes: "throughout legal history, each successive extension of rights to some new entity has been, theretofore, a bit unthinkable. We are inclined to suppose the rightlessness of rightless 'things' to be a decree of Nature, not a legal convention acting in support of some status quo … [that is why] the United States Supreme Court could straight-facedly tell us in *Dred Scott* that Blacks had been denied the rights of citizenship as a subordinate and inferior class of beings, who had been subjugated by the dominant race" (1972, p. 453). We have, however, thanks to Stone, since seen cases in which legal personhood has been successfully extended to environmental entities such as the Ganges River and sections of the Amazonian rain forest (Boyd 2018; Maloney 2018). In the case of the Ganges, the High Court in the northern Indian state of Uttarakhand in assigning such status determined that the river would be afforded all the "corresponding rights, duties and liabilities of a living person" including protection from harm (Margil 2017, p. 1). Clearly, an environmental entity cannot appear in court to defend these rights itself, however, as a corporate rights holder can employ legal representatives to bring prosecutions (for example against polluters) on its behalf.

The extension of legal personhood to such entities stretched conceptions of what constitutes a legal personality, sparking intense debates about where the limits of legal personhood could or should lie (O'Donnell 2017; Staker 2017). In an ongoing case in the United States, legal advocates have recently argued that two research chimpanzees Tommy and Kiko should be granted rights of *habeas corpus* (bodily autonomy) as a means of challenging their detainment (Coates 2017).

This would result in an acknowledgment (profound and far reaching) that the concept of legal personhood with attendant rights of liberty and protection could be extended to nonhuman beings. Although this argument was recently rejected in the State of New York Court of Appeals one of the presiding judges Eugene Fahey[1] expressed his consternation that the chimps had been denied "legal personhood" on the grounds that they could not "enter into the social contract" when it was clearly evident that if this were the only determinant of entitlement then members of the human species such as infants, comatose or demented persons would be similarly denied such rights. As Fahey concludes, the question of whether such entities should be treated as a person or as property, in essence a "thing," should not be taken as a simple either/or proposition. As he notes, the "evolving nature of life" ought to make such determinations the subject of ongoing discussion and debate.

I am interested in taking up Stone and Fahey's agenda of progressively reimagining how concepts of legal personhood and personality could, or should, be deployed by querying how they might be emplaced and constituted in the techne-mediated post-human condition and, thus, located and represented in law. I take as entry points to this discussion three case studies—one on bioinformation, one on novel neuro-regenerative devices and one on oocytes. I use these as vehicles to examine the new forms of virtual personality or personhood that are being brought into existence by developments in bioscience and the challenges they pose for key legal concepts such as dominion; liability or culpability, and identity. If "personhood" is viewed as the mechanism by which personality is to be protected it is worth first examining what, sociologically speaking, personality is generally thought to be and do, so that we might better grasp how it is morphing under conditions of post-humanism and what the legal and social implications of this evolution might be.

Personality according to the Oxford Dictionary can be understood as "the combination of characteristics or qualities that form an individual's distinctive character,"[2] an expression of personhood to which certain rights and obligations are due. The fact that they *are* due rests in part on a presumption that these "characteristics qualities," and I would argue the actions to which they give rise, are a consistent expression of the entity (either the legal or natural person) from which they originate. The difficulties that I want to draw attention to here are those that arise when either: (a) the qualities, characteristics, or actions of such entities can no longer be considered to be an *authentic representation* of that entity (legal or natural person); (b) are no longer under the unmediated control of that legal or natural person; or (c) are of such obscure origin, that it cannot be readily determined what those personalities, characteristics, or qualities are an expression of; thus complicating how legal or natural personhood can be traced and established in law. In seeking to answer these questions I address three key thematic concerns: dominion—who has sovereignty or control over the constructed personality; liability or culpability—who is responsible in law for the actions of technologized personalities; and identity—in this case, how genealogy affects the legal standing of prospective, though yet to be realized, persons.

Algorithmically Defined Selves: The Rise of the Data Revenant

Let us turn firstly to consider the emergence of a new iteration of personhood that I refer to here as the "algorithmically defined self" (Parry and Greenhough 2017, p. 146). We are all aware that historically large data sets have been employed to reveal population-wide patterns of health and disease. The census, for example, has been used to reveal patterns and rates of mortality, hospital and medical records to reveal rates or causation of disease. It was, however, very difficult to derive detailed information about individual persons from these datasets as their bioinformation was usually anonymized at the point of accession. This has recently changed as many data analytics companies are now able to combine identifiable personal information on health-related behaviors garnered from a variety of sources (Weber et al. 2014; Ng et al. 2015; Prainsack 2015). These include online social media platforms and forums, patient records, genetic profiles derived from genealogy sites; and large-scale databases that provide information on hospital visitations. Together they can be combined to create highly granular descriptors of how different groups of people experience health and disease.

Big data analysts have been encouraged to draw conclusions about the behavior of individuals and arrive at assessments of their risk through analysis of these kinds of metadata. As Viktor Mayer-Schönberger and Kenneth Cukier (2013, p. 12) put it: "just as Amazon can use such analytics to recommend books or Google to rank the most pertinent websites so too are they now used in diagnosing illnesses, recommending treatments and even identifying 'criminals' before they actually commit a crime." These kinds of predictive analytics are used to "algorithmically brand" particular individuals who appear, statistically at least, to meet the criteria for inclusion even if, in reality, the specifics of their case complicate this narrative.

Large organizations and corporations are drawn to such methodologies as they allow the complexity and chaos of our lived experiences to be reduced to data sets that can be interrogated algorithmically. Computer models determine the *probability* that individuals will perform in certain ways. They are in effect, creating a "proxy" for one's perceived personality, and its associated characteristics, qualities and most significantly predicted behaviors and actions. A new class of data doppelgangers is being generated by stitching together fragmented information about patterns of consumption and behavior from a variety of sources to create new digital personas and iterations of what Louveris (2014, p. 1) calls "digital personhood." One difficulty is that these revenant personalities may mirror but do not map perfectly to the identities we understand ourselves to inhabit, yet they nevertheless rise up to take on lives of their own.

I say this because, increasingly, it is these digital personas rather than the actual individual that are classified, assessed, and judged, and serious repercussions cascade out of this that can have lasting economic and social implications and discriminatory effects. Decisions that will dramatically affect all of our lives: from whether we will be given an expensive new life-saving drug or be allowed to move into better social housing will increasingly rely not on how we behave but rather on how our algorithmically defined selves are predicted to behave.

The result is a world in which the state and other authorities intervene into our lives (or oblige us to make changes in our lifestyle) based not on who we are, but on whom our data makes us appear to be. These data doppelgangers are not, it could be argued, a consistent or authentic expression of the natural person from which they arise, but rather take on a spectral quality. Mary Ebeling (2016, p. 7) describes these constructed algorithmic personas as "data revenants"—"they are not ghosts, they have bodies, but hollow bodies, puppet like shells ... that return to inhabit an uncanny life." In a multi-platform world, many iterations of a person's identity may emerge and multiple digital personas may coexist at any given moment. Conventionally, an individual is understood socially, and in law, to be a single person. The question thus arises: how will it be possible for a live individual (a natural person) to manage other's perceptions or prevent them from assessing or judging these digital or algorithmic personas as if they were completely consistent and reliable proxies for the person themselves?

This governance issue matters because data revenants are now becoming vitally important economic subjects, albeit alteric ones. Organizations and corporations now make decisions on key matters such as the imagined trustworthiness of individuals, and the utility or value of interventions pertaining to them, based on assessments of the character and perceived behavior of the data revenant, even if that is typically constructed out of a collation of partial, inaccurate, or incomplete sets of derived bioinformation. Such practices can have seriously discriminatory effects. Consider, for example, the single parent denied a mortgage because their data suggests they are too great a financial risk, or the retired couple refused health treatment because their chances of recovery are deemed to be statistically slim.

Although these scenarios might seem fantastical, practices such as "health risk scoring," in which personal data from sources as disparate as social media profiles and hospital attendance is aggregated to model an individual's likely risk of becoming ill, are becoming much more commonplace. Their applied use has certainly risen exponentially in the United States since the 2007 implementation of the Affordable Care Act which compensates insurers based on patient risk. Many questions have been raised about the ethics and the accuracy of making probabilistic assumptions regarding the behavior of individuals or their health status, from such information. Whist these predictions may prove correct in some instances, the history of actuarial practice reveals that even those predictive models that attend to much simpler matters, such as general life expectancy (Ayuso et al. 2018), can fail to produce reliable estimations, as evidenced, for example in current shortfalls in the UK's pension funding provisions.

The attraction of using social media data to inform assessments of risk in the insurance industry was demonstrated recently in the UK when the car insurer Admiral unveiled its new analytic tool for mining individual's Facebook posts for "personality traits linked to safe driving," using these to price insurance offerings. Those who write in short, concise sentences, use lists, and arrange to meet friends at a set time and place, rather than just "tonight," were identified as conscientious, in contrast to those whose more excitable use of exclamation marks and phrases such as "always" or "never" rather than "maybe" led them to categorized as "overconfident." Under the scheme the former were determined to be "lower risk" and offered discounts of up to £350 on their premiums per year.[3]

Judgments such as these, that are now determining access to health care, likely predisposition to disease, or rights to access insurance are increasingly performed computationally by software that acts as an autonomous decision-maker. A further case illustrates how ethically problematic this can be. Identifying those at risk of mental illnesses such as depression or alcoholism has always proven complex. Individuals are loathe to divulge their risk to health care practitioners as they fear wider societal discrimination that might hamper their chances of securing employment or insurance. Psychiatric researchers are now suggesting that one way of getting around this problem of "non-disclosure" would be to "mine" online social media posts for information on risky behavior that might improve the "detection" of those with serious mental health conditions. This online information, which includes, tweets, status updates, and images, is, such researchers suggest (Inkster et al. 2016, p. 1089), more reliable than offline self-reported information as it "reflects valid depictions of offline behaviors (displays of alcoholic behaviors that indicate actual misuse); measures content that is difficult to assess offline (for example, conversational intensity); achieves previously inconceivable sample sizes; and is more cost effective." The adoption of this methodology is being proposed despite the fact that disclosures made on social media may be exaggerated or false, and thus lack clinical diagnostic validity.

What is being "scored" through such exercises, I would argue, is not the natural person but rather the constructed data revenant, that has risen up to take on a personality of its own, even if that personality proves not to be a very accurate iteration of the individual from whom it was derived. The ethical implications of allowing some members of society to be abandoned as "untenable" in this way are considerable but also compounded by the fact that many individuals remain unaware that they have attracted a negative score until their insurance premiums rise or they are declined treatment or employment. Lack of public awareness of the role of algorithmic techniques along with the facelessness of the Big Data industry combine to obscure the fact that these individuals are increasingly being substituted for by data revenants: new alteric digital subjects that might perhaps be best thought of as the uncanny progeny of highly entwined and interdependent human–machine interactions.

Disentangling who or what is responsible for their creation, of who has dominion over them or who might be considered liable for the inaccurate decisions they make (to whom, or to what, appeals against such decisions could or should be directed) remains, at present, largely unresolved. The "actant" that is the autonomous algorithmic software package both generates and then adjudges the riskiness of the revenant personality but can make inaccurate assessments. At present it is primarily assumed in law that the software program is the agent of an identifiable human or collective of humans, in other words that it does not have a legally recognized personality of its own. It is not considered to have the ability to act in its own right, or to be held responsible for its actions. This, however, creates a liability gap.

The European Parliament, in attempting to address this issue, decided in January 2017 to recommend establishing a special legal status for robots to be recognized as "electronic persons" that is to say, as legal entities with special rights and obligations that could include redress for any damage they cause (European Parliament Resolution on Civil Law Rules on Robotics Section 59 (f)).[4] The question of whether such software is genuinely autonomous or capable of acting with intentionality is, of course, immensely complex in and of itself. Teubner's elegant typology provides a useful tool for undertaking such assessments. As he argues: "software agents act autonomously in the legal sense when their behavior no longer follows an 'exclusively stimulus-reaction scheme'[5] but when they pursue their own goals and make decisions that cannot be predicted" (Teubner 2018, p. 14). The task of locating not only autonomy but liability (or should we say culpability?) when wrongful decisions are made is difficult, but further complicated in instances when decision-making is produced, as it increasingly is now, out of complex networking between multiple computer interfaces.

Revenant personalities are constructed by linking together information held on a variety of computer network interfaces from medical record databases to social media platforms. Each of these has, in principle, a custodian who is responsible for the ways in which they are used. What happens though when the persona only emerges out of the traffic between these interfaces? Who should then be considered to be their "author" or "owner"? As the legal scholar Bryant Smith noted as long ago as 1928, the solution to the question of culpability in such circumstances can be found in the device of "perpetual succession" "by virtue of which each shareholder becomes still less significant, and even presidents and boards of directors lose their identity in the regular flow of successors." Those who wish to press charges or make claims are thus directed to a faceless organization or entity—such as the e-person that becomes the responsible "legal personality." So it is, Smith suggests, that "for one purpose and another, it becomes convenient, if not indeed necessary, to let the individual participants fade out of the picture and to look upon the organization as the unit ... [that can be] treated and spoken of as if it were a natural person" (Smith 1928, pp. 288–289). Should we allow data custodians or shareholders to thus fade from view, or should each take a more prominent and accountable role as "shareholder" in the juridical fiction that is the "legal personality"? Does either approach adequately protect social interests in the twenty-first century? These, I would suggest, remain matters worthy of further consideration.

Abolishing the Brain/Machine Interface

The concept of a human-digital-technological association or interface is also a useful means by which to conceptualize other kinds of hybrid entities that are now emerging in the biomedical realm which pose similar challenges to concepts of personality and personhood. These take the form of new biomedical devices that are designed for use within human beings. As I have noted elsewhere, novel combinatorial devices are now being developed for the remediation of complex conditions that are constituted of an amalgam of human, animal, and synthetic materials designed to become imbricated into the human body and therefore, ontologically, an indistinguishable part of it (Parry 2018).

Historically, devices have been presumed to be materially inert rather than lively; objects that can be as easily extracted from the body as they are implanted. Little attention has yet been given to the political agency of these new devices or the question of how they might act to alter, in fundamental and potentially irreversible ways the personality of their proprietors. It is useful here to take as an entry point to this discussion the case of novel neuro-regenerative devices and therapies that are intended to operate in ways that supersede historical understandings of the device as purely mechanical in operation. These new devices are designed, alternately, to become, in effect, active participants or interlocutors in a "conversation" between this technology and the person they are being inserted into, a conversation that can radically alter what or whom that person will become.

As Meloni et al. (2011) explain, devices such as brain implants and interventions such as brain gene transfers have been developed to restore neural functions for therapeutic purposes, but also have the capacity to be externally monitored, controlled and even modified. Brain micro-implants act as sensors and signal transductors to restore visual and hearing loss (Bishop and Eby 2010; Chader et al. 2009; Fletcher and Greferath 2010), whilst output signal transducers are designed to help bridge the motor cortex and peripheral limbs to restore function following severance of the spinal cord (Capogrosso et al. 2016). Those that are likely to have the most impact on personhood and personality are stimulating electrodes that are implanted into specific structures located inside the brain. This form of deep brain stimulation (DBS) is now used as a treatment to correct essential tremor found in Parkinson's disease but also to treat psychiatric conditions such as depression (Dougherty et al. 2015) and has even been promoted as a means to correct behavioral traits such as addiction (Salling and Martinez 2016).

The presumption that an individual's personality and associated actions are a product of their unique neurological composition is eroded by the introduction of devices such as these that have a recognized capacity to, as Jane Bennett describes it, "exceed their status as objects and to manifest traces of independence or aliveness" (2009, p. xvi). Indeed, it could be argued that in some instances, the device begins to operate autonomously within the recipient body, escaping to colonize it in ways that induce behavior that can no longer be considered to constitute an authentic expression of that individual's former personality and therefore their natural personhood in law. Scientific studies on the impact of DBS on Parkinson's patients have shown significant changes in personality traits in some individuals after just three months use. The most significant of these findings is that some patients showed increased impulsivity, including tendencies to make hasty choices, leading researchers to hypothesize that DBS may interfere with their ability to hesitate and to weigh options when facing difficult decisions. According to the observations of close relatives, these patients were considered to be less thoughtful, more impulsive, and more likely to act on the spur of the moment without regarding the consequences after DBS, than they were before (Pham et al. 2015).

In this sense, it could be argued that decision-making is no longer fully under the control of the natural person as they have experienced a fundamental and potentially irreversible alteration to their preexisting personality, which has become virtual in accordance with the definition "almost or nearly as described, but not completely or according to strict definition." Such devices have the capacity as Meloni et al. (2011, p. 1) succinctly put it, "to abolish the brain/machine interface" and, in so doing, to complicate historical notions of legal responsibility by acquiring their own personalities (through subversion of their owners'), as they redirect thought and action through their reflexive engagements with the complex architectural milieu of the post-human body. The question of where, to whom or what, to attribute culpability or liability in circumstances where these mediated actions induce "tortuous" harms, remains, as yet, largely unexplored in both sociological and legal terms.[6] Who or what would be considered culpable in law for the actions of these technologized personalities?

Spectral Personas: Origins and Identity

If, by definition, we understand a personality to be constituted of the combination of characteristics or qualities that form an individual's distinctive character, then curiosity inevitably leads us to wonder from whence those characteristics and qualities derive. Personality is a construction, I would argue that cannot be understood outwith consideration of questions of genealogy and must therefore sit within the broad temporal arc of preconception, conception, birth, lived experience, death, and afterlife. As we know, for example, the concept of personality rights, in International Property Rights law (IPR) reflects the right of an individual to control the commercial use of his or her name, image, likeness, or what are considered to constitute other unequivocal aspects of their identity. It is interesting to note that personality rights are primarily understood to constitute a property right (rather than a private right) thus they remain actionable even after the death of the natural person concerned. So, for example, family or charitable trusts can sue for damages if they believe that the privacy of a deceased relative has been invaded or if the quality of their (former) personal reputation has been unfairly besmirched through publication, broadcast, or other means. In this sense, these personality rights (which pertain closely to the transmission and protection of one's inviolable identity) are understood as temporally extensive, if not, even perhaps, *heritable*.

The findings of fieldwork that I and others, such as Catherine Waldby (2019) have recently conducted on the global oocyte economy has revealed how the genealogy of an individual's biological heritage can become permanently obscured through the black-boxing of the derivation of their genetic material. In many of the reproductive hubs of the global assisted reproduction industry (Mumbai, Bangkok; Pnom Penh, and Bucharest) young women sell their eggs to clinics to service the needs of infertile intending parents. These women are frequently encouraged to participate in the market through the creation of savior narratives in which they rescue a named (but not identified) couple from the ignominy of infertility through the gifting of their gametes. What they are not told, and thus do not realize, is that the levels of ovarian stimulation to which they are subjected will lead to the extraction of as many as twenty-five eggs in a single cycle. Whilst some of these will go to the named intending parents, the others will enter a wider resource pool of gametes that the clinic may use for their other clients or circulate to consumers in other branches of their clinics elsewhere or beyond. The gametes become through this process an asset of the reproductive corporation, part of its property portfolio.

It is often assumed in these contexts that the progeny generated from these donated gametes will either remain unaware of their derivation, or alternately, be so ashamed of it (as donors are often from lower castes and classes than commissioning parents) that they will never seek to determine who their egg donor was, or to meet them. However, the western obsession with genealogy, what the Guardian journalist Zoe Williams describes as "an exercise in self-congratulation"[7] has become a much-favored pastime as people seek to establish what traits, personal characteristics or qualities they may share with their antecedents or how their own life experiences may be differentiated (usually improved) from that of their ancestors through processes of social mobility. So great has been the drive to determine how and in what ways our identities—and one may argue personalities—sync with, or diverge from, that of our ancestors that interest in genealogy research has exploded, garnering an estimated forty million practitioners in the United States alone by 2004 (Tallbear, cited in Rak 2017). That the children of a globally extensive fertility market would not seek to find their donors thus seems highly unlikely to say the least.

However, in the rush to secure the services of donors, and in the absence of awareness that donor-conceived children would wish to know their biological or genetic heritage little attention has been paid to accurately recording the source or destination of oocytes derived from these emerging economies. The fact that such gametes may be cryogenically preserved for hundreds of years before use, stretches the temporality of this disassociation even further. Together, these technological advances have served to thoroughly obscure the genealogy of the identity and personality of an entire constituency of persons so conceived. This potentially creates quite severe transgressions of human rights for those who will remain forever unable to establish, or establish with any certainty, from whom their personal qualities and characteristics derive. The genesis of their prospective personhood and personality becomes, as a consequence, almost, in this sense, spectral.

The UN Convention on the Rights of the Child stresses the value and importance of a child's identity. Under Article 7, a child has a right to have a "legal" identity: rights to be registered, to have a name, and a nationality. Article 8 also explicates the States' responsibility to the protection of more dynamic elements of his or her identity. This includes as Giroux and De Lorenzi (2011, p. 60) suggest "morals and religious and cultural characteristics" that are transmitted through "family life." In a leading human rights case *Gelman v Uruguay*, the Courts invoked Articles 3 (Right to Juridical Personality), 4 (Right to Life), 5(1) (Right to Physical, Mental, and Moral Integrity), 7 (Right to Personal Liberty), 17 (Rights of the Family), 18 (Right to a Name and to Surname of Parents), and 19 (Rights of the Child) of the Convention to explicate how the rights of the child of one of Argentina's political "disappeared" had been abrogated by her removal from her biological parents and placement within an adoptive family of Uruguayan extraction.

In adjudging the case the Court noted that "learning the truth of her parentage has profoundly changed her life." In finding for her, it noted that the State had effectively made it impossible for her to know her family or to establish relations with them. By these actions the State had effectively "suppressed her identity," "deprived her of her true nationality by preventing her birth in Argentina" and most significantly, "deprived her of juridical personality." The Court further found that her separation from her biological mother constituted a violation of personal liberty because "in the context of childhood, such a separation inherently violates a child's liberty." Interestingly the Court also noted that the violations of her rights as enshrined in the Convention occurred within the context of her "right to identity," which the Court defined as "the collection of attributes and characteristics that allow for the individualization of the person in a society" (Tripodes 2015, pp. 1874–1876)

The resonances here with the question of donor-conceived children who are the product of undocumented donations are clear and present. These children form, in a sense, a new population of technologically generated assemblages of uncertain origin and identity. Is it important that they have a formal right to establish their own identities beyond doubt? Whilst some scholars, such as Bently (2010) believe that enshrining personal identity in law could restrict an individual's opportunity to freely construct their own personality in their own lifetime, others believe that the application of human rights law should enable individuals to excavate key genealogical aspects of their identity that have been deliberately and unlawfully obscured from their view. The ability to access such information, which allows the genesis of their prospective personhood and personality to be discerned, may well prove for many to be a vital actuator of their own personal identity.

Conclusion

The cases that I have outlined here, though quite divergent, together demonstrate how historical understandings of the generation and transmission of personality are being disrupted by technological change. How are, or should, concepts such as dominion, culpability, or identity be sustained or nuanced in response? In this chapter, I have begun the task of exploring where the genesis of personality lies—here we need to pay attention to the concepts of derivation and authenticity. Something needs unpacking here about the idea of personality formation. The presumption that personality is a constant and authentic expression of a natural person is being collapsed by developments in the contemporary biosciences in the ways I have described above. Multiply constituted revenant or spectral personas may emerge as viable representations of the natural persons from whom they derive, even if they prove not to be reliable or consistently "authentic" ones.

In these cases, we have seen how revenant personalities (data doppelgangers or engineered personas) can rise up to take on lives of their own—and be adjudged as reliable proxies for the natural person. The natural person can even be held lawfully accountable for their actions. This raises very serious questions about the limits of the liability and culpability. What is the ontological standing of a data revenant in law? Should they be considered the "property" of the natural person from who they are derived, or alternatively, be understood as a piece of constructed intellectual property—an artificial person and legal personality that belongs to the data analysts that develop them? Who should consequently be understood to have sovereignty or dominion over these new legal personalities? What of the mutated personalities that find expression when the brains of their natural owners are technologically stimulated in unprecedented ways? What or who should be held accountable for their actions?

Being prepared (following Stone's example) to "think the unthinkable" and consider from whence or what personality derives and how this might be queered or obscured by technological intervention should also prompt us to begin the vital work of considering how the rights and obligations of virtual or spectral personas might be observed or prosecuted in the future. Excavating the genealogy of personhood—querying whether it arises from the fabric of the person, or through a longer engagement with their biological antecedents; whether it constitutes a consistent or authentic expression of selfhood and examining how and under what conditions this might change, will provide much needed opportunities to reflexively explore both our understandings of personality and how and in what ways it might find rightful expression in legal personhood in a post-human age.

Notes

1. The full text of Fahey's judgment is available at https://www.nycourts.gov/ctapps/Decisions/2018/May18/M2018-268opn18-Decision.pdf.
2. https://en.oxforddictionaries.com/definition/personality.
3. https://www.theguardian.com/technology/2016/nov/02/admiral-to-price-car-insurance-based-on-facebook-posts.
4. European Parliament resolution of 16 February 2017 with recommendations to the Commission on Civil Law Rules on Robotics (2015/2103(INL) Section 59 (f)), available at http://www.europarl.europa.eu/sides/getDoc.do?type=TA&language=EN&reference=P8-TA-2017-0051.
5. E.g. Interrogate a database and receive a response.
6. Although see Ovadia, D & Bottini, G 2015, 'Neuroethical implications of deep brain stimulation in degenerative disorders', *Current Opinion in Neurology*, vol. 28, no. 6, pp. 598–603.
7. Williams, Z 2009, 'The story of you, yes you!', *The Guardian*, 14 January 2009, available at https://www.theguardian.com/commentisfree/2009/jan/14/social-mobility-genealogy-census-website.

References

Ayuso, M, Bravo, J, Holzmann, R 2018, *Getting life expectancy estimates right for pension policy: period versus cohort approach*, Discussion paper series IZA Institute of Labour Economics, available at http://ftp.iza.org/dp11512.pdf.

Bennett, J 2009, *Vibrant matter: a political ecology of things.* Duke University Press, Durham.

Bently, L 2010, 'Identity and the law', in G Walker & E Leedham-Green (eds), *Identity: the Darwin College lecture series.* Cambridge University Press, Cambridge.

Bishop, C & Eby, T 2010, 'The current status of audiologic rehabilitation for profound unilateral sensorineural hearing loss', *The Laryngoscope*, vol. 120, no. 3, pp. 552–556.

Boyd, D 2018, 'Recognizing the rights of nature: lofty rhetoric or legal revolution?' *Natural Resources & Environment*, vol. 32, no. 4, pp. 13–17.

Capogrosso, M, Milekovic, T, Borton, D, Wagner, F, Moraud, E, Mignardot, J et al. 2016, 'A brain–spine interface alleviating gait deficits after spinal cord injury in primates', *Nature*, vol. 539, no. 7628, p. 284.

Chader, G, Weiland, J, Humayun, M, Verhaagen, J, Hol, E, Huitenga, I et al. 2009, 'Artificial vision: needs, functioning, and testing of a retinal electronic prosthesis', *Progress in Brain Research*, vol. 175, pp. 317–332.

Coates, A 2017, 'Is this Chimpanzee a non-human person?', *The Independent*, 13 September, https://www.independent.co.uk/news/long_reads/is-this-chimpanzee-a-non-human-person-a7941876.html.

Dougherty, D, Rezai, A, Carpenter, L, Howland, R, Bhati, M, O'Reardon, J et al. 2015, 'A randomized sham-controlled trial of deep brain stimulation of the ventral capsule/ventral striatum for chronic treatment-resistant depression', *Biological Psychiatry*, vol. 78, no. 4, pp. 240–248.

Ebeling, M 2016, *Healthcare and big data: digital specters and phantom objects.* Palgrave Macmillan, Houndmills.

Fletcher, E & Greferath, U 2010, 'Electronic restoration of vision: science fiction or reality?', *Clinical and Experimental Optometry*, vol. 93, no. 2, pp. 59–60.

Giroux, M & De Lorenzi, M 2011, 'Putting the child first: a necessary step in the recognition of the right to identity', *Canadian Journal of Family Law*, vol. 27, pp. 53–94.

Inkster, B, Stillwell, D, Kosinski, M & Jones, P 2016, 'A decade into Facebook: where is psychiatry in the digital age?', *Lancet Psychiatry*, vol. 3, no. 11, pp. 1087–1090, available at https://doi.org/10.1016/S2215-0366(16)30041-4.

Louveris, P 2014, https://gtr.ukri.org/projects?ref=EP%2FL005859%2F1.

Maloney, M 2018, 'Environmental law: changing the legal status of nature: recent developments and future possibilities', *LSJ: Law Society of NSW Journal*, vol. 49, p. 78.

Margil, M 2017, 'India court declares "personhood" of glaciers and ecosystems', Press Statement, 3 April, http://celdf.org/2017/04/pr-india-court-declares-personhood-glaciers-ecosystems/, 27 April 2017.

Mayer-Schönberger, V & Cukier, K 2013, *Big data: a revolution that will transform how we live, work and think*. John Murray, London.

Meloni, R, Mallet, J & Faucon Biguet, N 2011, 'Brain gene transfer and brain implants', *Studies in Ethics, Law, and Technology*, vol. 4, no. 3, pp. 1–17, https://doi.org/10.2202/1941-6008.1141.

Ng, Kenney et al. 2015, 'Personalized predictive modeling and risk factor identification using patient similarity', *AMIJoint Summits on Translational Science proceedings. AMIJoint Summits on Translational Science*, vol. 132, no. 6, p. 2.

O'Donnell, E 2017, 'At the intersection of the sacred and the legal: rights for nature in Uttarakhand, India', *Journal of Environmental Law*, vol. 30, no. 1, pp. 135–144.

Ovadia, D & Bottini, G 2015, 'Neuroethical implications of deep brain stimulation in degenerative disorders', *Current opinion in neurology*, vol. 28, no. 6, pp. 598–603.

Parry, B 2018, 'The social life of "scaffolds" examining human rights in regenerative medicine', *Science, Technology, & Human Values*, vol. 43, no. 1, pp. 95–120.

Parry, B & Greenhough, B 2017, *Bioinformation*. Polity Press, London.

Pham, U, Solbakk, A, Skogseid, I, Toft, M, Pripp, A, Konglund, A et al. 2015, 'Personality changes after deep brain stimulation in Parkinson's disease', *Parkinson's Disease*, pp. 1–8.

Prainsack, B 2015, 'Three H's for health—the darker side of big data', *Bioethica Forum*, vol. 8, no. 2, pp. 40–41.

Rak, J 2017, 'Radical connections: genealogy, small lives, big data', *a/b: Auto/Biography Studies*, vol. 32, no. 3, pp. 479–497.

Salling, M & Martinez, D 2016, 'Brain stimulation in addiction', *Neuropsychopharmacology*, vol. 41, no. 12, pp. 2798–2809.

Smith, B 1928, 'Legal personality', *Yale Law Journal*, vol. 37, no. 3, pp. 283–299, available at http://digitalcommons.law.yale.edu/ylj/vol37/iss3/1.

Staker, A 2017, 'Should chimpanzees have standing? The case for pursuing legal personhood for non-human animals', *Transnational Environmental Law*, vol. 6, no. 3, pp. 485–507.

Stone, C 1972, 'Should trees have standing? Toward legal rights for natural objects', *Southern California Law Review*, vol. 45, pp. 450–501.

Teubner, G 2018, *Digital personhood? The status of autonomous software agents in private law*, available at SSRN https://ssrn.com/abstract=3177096 or http://dx.doi.org/10.2139/ssrn.3177096.

Tripodes, Z 2015, 'Gelman v. Uruguay Loyola Los Angeles', *International and Comparative Law Review*, vol. 37, pp. 1861–1885.

Waldby, C 2019, *The oocyte economy: the changing meaning of human eggs*. Duke University Press, Durham.

Weber, G, Mandl, K & Kohane, I 2014, 'Finding the missing link for big biomedical data', *The Journal of the American Medical Association*, vol. 331, no. 24, pp. 2479–2480.

CHAPTER 3

The Political Economy of Neurolaw: Can Neurolaw Destabilize the Neoliberal Discourse About Human Behavior?

Delphine Rabet

Introduction

Ideas and representations matter because they have material consequences. Neuroscience is a rapidly developing field that offers new ways to think about human behavior, in particular around notions of responsibility and accountability; the law has been relatively quick to engage with neuroscience and this has resulted in novel representations of the human experience through the field of neurolaw and subsequently challenges to the well-established concept of legal personhood. This process starts at the ideational level and then gets translated into practices developed inside the courtroom but also applicable to the broader society.

Legal personhood relies on a particular representation of what it means to be a person according to the law. This representation is centered on concepts of rationality, reasonableness, choice, intentionality, and responsibility among others. However, neuroscience may question

D. Rabet (✉)
University of New South Wales, Sydney, NSW, Australia
e-mail: d.rabet@student.unsw.edu.au

M. de Leeuw and S. van Wichelen (eds.),
Personhood in the Age of Biolegality, Biolegalities,
https://doi.org/10.1007/978-3-030-27848-9_3

the core aspects of "personhood" and by extension of legal personhood by documenting the mechanics of the decision-making process and ultimately showing that each of these criteria of personhood is biologically contentious at best. Destabilizing the concept of legal personhood in this way would have, as a flowing effect, to require new conceptualization of guilt and innocence and some reshaping of the justice system.

This interaction between law and neuroscience is therefore worth studying but needs to be understood as an evolution occurring within a larger ideological moment, namely, Neoliberalism. This chapter argues that if at first glance neurolaw seems to fit in the contemporary neoliberal ethos, unsurprisingly, it also offers sites of resistance and contestation that challenge the neoliberal narrative. However, further analysis of the knowledge coproduced by neuroscience and the law shows that this challenge is mostly superficial and ultimately, neurolaw may instead consolidate a representation of the self best normatively described by Neoliberalism.

Neuroscience, Neoliberalism and the Law: Sites of Knowledge Coproduction

This first part argues that the way in which neoliberalism, neuroscience and the law intersect to produce new knowledge about legal personhood creates a particular representation of human beings centered on the dominance of individual responsibility and accountability as both an ontological claim and a normative project. It starts by discussing separately neoliberalism and neuroscience and then focuses on the kind of knowledge coproduction occurring through legal processes.

Neoliberalism has become one of these contentious concepts vastly overused in some academic circles and yet almost unknown to most of the general population. Still, it presents some analytical tools that are useful to better understand the dynamics at play in the current phase of late capitalism. In simple terms, neoliberalism is both an ideology and a set of practices that see the individual as inherently rational, self-interested, and able to express freely these interests and preferences through the choices he or she makes. If Marxist scholars tend to emphasize that neoliberalism is an expression of class conflict and focus on the changing relationship between the state and society due to the dominance and expansion of Capital (Harman 2008), a Foucauldian perspective seems more appropriate for the purpose of this chapter.

Although Foucault can be ambivalent about the advantages and pitfalls of the coming of a neoliberal self (Zamora 2014), his analysis highlights which kind of representations of the self-results from the neoliberal project. These new representations are linked to the dominance of a particular economic discourse centered on the subjective preferences and conduct of the individual—or *homo economicus*—and expand a market logic to all social and political relations (Harvey 2005). In doing so, neoliberalism transforms the individual into an "entrepreneur of himself" (Foucault 2008, p. 226) who engages not only in exchanging but most crucially in competing with other human beings in every aspect of life (Read 2009). Through biopolitics, the individual is made aware of what represents his "human capital" (Becker 1964), in other words the capabilities, behavior, and knowledge that will make him more desirable on the market place. In fact, the person and his human capital are inseparable (Schultz 1972), it is in the biological self that this human capital is inscribed and it is therefore through the body that a particular type of governmentality must take place.

Homo economicus does not need to be directly governed however but instead internalizes the expectations of the market and responds to changes and new demands based on the information available to him. This shift in the way individuals see themselves and are seen as subjects has material consequences in terms of the level of responsibility they are expected to carry as the result of their conduct in every aspect of their lives. Getting the correct information about one's biological self becomes a key tool to maximize one's success in economic and other types of markets; developments in neuroscience participate in achieving this goal.

Most neuroscientific research claims that human behavior can be traced back to observable neurological processes located in the brain (Pickersgill 2013). Going even further and coming from a critical perspective, Vidal argues that "brainhood" is in fact replacing personhood and that therefore, "we are our brains" (Vidal 2009, p. 6). Individuals have become neurochemical or "neuronal selves" (Pitts-Taylor 2010, p. 639) who understand their "thoughts, feeling and behavior as being mediated through the brain" (Rose 2007, p. 188). This neurocentrism is at the heart of neuroscience, reducing the human experience to cerebral activities.

Similarly, neoliberalism suffers from a reductionism that considers that human beings are free agents, animated by a market logic that extends outside of the economic sphere to the private and emotional dimensions of life. *Homo economicus* is therefore only defined by how much and what kind of human capital he can invest and maximize when he is engaging in human activities. This resonates with the attempt from neuroscience to only consider how much and what kind of brain activity is occurring to explain human behavior.

Another parallel between neoliberalism and neuroscience exists in terms of the characteristics associated with both markets and brains. Markets on the one hand and brains on the other are conceived as "natural" and nonhierarchical organization populated by networks and where flexibility, adaptability to new circumstances and demands are the norm (Slaby 2010). The self-organizing ability of markets and brains seem to constitute a clear connection between a neoliberal political economy and neuroscience.

Investing in human capital, within a neoliberal framework, represents the ultimate responsibility of the individual. Developing intellectual skills leading to a better income, maintaining a healthy body or migrating to a country where new opportunities may arise all represent important investments in human capital (Read 2009). There are some obvious biological limitations though to these investments as physical and mental capacities are partly determined by genetic factors. Neuroscience represents a perfect option to access information and address biological limitations constraining individual behavior in economic, social, or legal spheres. This new knowledge of the self produced by the interaction of neuroscience and legal processes is the focus of the next section.

Scientific and legal knowledge are never produced in a vacuum. Instead, any form of new knowledge is closely related to the broader political economy structure: its dominant ideas, institutions, narratives, and representations. As developed above, neuroscience—and biotechnologies in general—fits in the current neoliberal representation of the self by focusing exclusively on individual behavior and taking it at face value. This process obscures at the same time the importance of the ecological conditions, understood broadly, necessary to the development of human beings (Parry 2018). Neuroscientific knowledge then validates the neoliberal view of atomized individuals best illustrated by Margaret Thatcher when she claimed "there is no such thing as society" (Thatcher 1987), only individuals.

Legal knowledge already accommodates a neoliberal ethos by emphasizing and promoting the existence of a neutral free legal subject, the sanctity of contract, or the unchallengeable concept of private property. In more recent times, the law has pushed the boundaries of knowledge creation even further by providing the legal tools to commodify and privatize the living to levels never achieved before. The new intellectual property rights regime best illustrates such a development (May 2009). It is on these fertile grounds that knowledge coproduction occurs.

Science and the law have a long history of engagement as, for instance, expert scientists of various kinds, and increasingly neuroscientists, have offered their specific knowledge to solve or inform legal questions. This engagement has for effect to both consolidate the credibility of neuroscientists (Jasanoff 2007) as a new group of experts, and also to validate legal decisions because they appear supported by scientific facts.

It is however the idiom of coproduction which is most useful to understand how legal processes are partially built on scientific truths. More broadly any knowledge is embedded in social, political, and economic structures that become also receptive to the new knowledge created (Jasanoff 2004). Such an approach highlights the complex interaction between knowledge, culture, and power. It also illustrates how specific sites of knowledge production such as the laboratory or the courtroom reach new audiences and affect other areas of knowledge production. Sheila Jasanoff (2007, p. 772) is right to say that "As agents of power, law and science also collaborate in sustaining wider understandings of how society works, including ideas of the human self and agency, the market, and the collective good."

Neuroscientific facts are absorbed, reinterpreted by and included in legal processes that will allow reaching final decisions on legal questions. This happens in spite of the tension between scientific and legal methodological and epistemological positions. While science is a dynamic process with constant changes and developments due to new discoveries, the law is static and establishes processes that are difficult to contest, it needs certainty and definite answers to offer closure to the parties involved. Yet, paradoxically, the law will follow scientific reasoning, select certain scientific facts and integrate them into the legal process to reach legal conclusions.

Establishing "responsibility" is a key question for the law. The development of neuroscience and its inclusion in the legal process brings new information and requirements to meet this imperative.

Through imaging, neuroscience claims to identify if the biological aspect of the decision-making process located in the brain is considered as normal or abnormal (Wasserman and Johnston 2014). With this information, the law can then determine whether or not, legally, the accused can be labeled as "responsible." The conjunction of these two regimes of truths creates a particular representation of the notion of responsibility that is reduced to brain activity and neurological representations of the self, in other words, the law becomes only interested in the neuronal self (Pitts-Taylor 2010). What constitutes an individual as sane, and therefore responsible, or insane—either fully or partially—and therefore irresponsible or with limited responsibility, is not just a legal discourse but it is grounded in neuroscientific truths.

This shows how the coproduction of the legal process that seeks to establish responsibility is informed by both scientific and legal knowledge and results in a new representation of the individual, reduced to its biological identity and even further to its cerebral activity. These definitions of responsibility and insanity are in turn aligned with the neoliberal project in that they focus on a reality that belongs exclusively to the individual and tend to exclude or obscure other aspects relevant to an assessment of a person's responsibility.

Another illustration of the coproduction of knowledge occurring through neurolaw is the potential to create new categories of individuals by selectively defining the kind of knowledge relevant or not to the court (Barry et al. 2008). Following Ian Hacking's argument (Slaby 2010), combined scientific and legal knowledge may be able to "create" novel kinds of people. In either a criminal or civil case, neuroscientific evidence may provide specific information about brain dysfunction linked to excessive alcohol or drug consumption for instance. This would not only have an effect on the level of responsibility of the accused and most importantly could also portray the individual as a failed entrepreneur of himself for instance as a new "imagined" category of person.

Once again this perspective is aligned with neoliberalism because it segments a market of behaviors in new ways that would then lead to a new form of control, surveillance, and governance while at the same time creating new markets for the treatment of such conditions, previously thought of just "bad behavior."

In short, knowledge coproduced by the interaction of neuroscience and the law also intersects with neoliberalism and consolidates a particular representation of the human experience based on personal responsibility and an expansion of a particular kind of economic dogma to all other aspects of life. However, neoliberalism is a contested ideology and set of practices and there are a number of ways in which the knowledge coproduced by neuroscience and the law can appear to clash with the neoliberal project. The next section now turns to this question.

Neurolaw and Contesting the Neoliberal Narrative

In both neoliberalism and the law, there is an assumption that all individuals are equal in their ability to make choices that will reflect their preferences and that they do so by exercising their free will. Neuroscience however claims that particular human behaviors are linked not to moral decisions but to the shape of and the activity happening in the brain. The individual is unable to behave in any ways contrary to how he or she is biologically determined. In other words, the mind has no control over the brain, and the brain controls the body. Neuroimaging can actually confirm this determinism and visually illustrate neurological activity—or lack of—that could explain specific behavior (Wasserman and Johnston 2014). In fact, neuroscience questions the existence of any form of free will[1] and therefore undermines greatly the legal definition of responsibility (Dumit 2004), itself heavily influenced by what constitutes moral responsibility (Wasserman and Johnston 2014).

With this neurolegal knowledge, the notion of *homo economicus* gets similarly challenged when one transfers this truth to the political economy realm. Leaving aside compatibilist theories (Hogson 2000), the idea of a rational individual able to choose freely and therefore carrying full responsibility for their actions becomes untenable if their choices are biologically predetermined. In this instance, new knowledge does not consolidate but instead undermines the dominant representation of the neoliberal human experience.

Neurolaw may as well provide an opportunity to gain further understanding about the circumstances of an individual's life that resulted in a particular type of brain architecture. The law already considers the life of an accused when it comes to sentencing and new neurolegal knowledge

[1] The Libet experiment is one of the most well-known attempts to establish that individuals have no free will. It posits, based on brain activity imaging that there is a gap between the time one's brain decides to act and the moment one makes the conscious decision to act.

could go a step further. It could acknowledge, as first argued by Soviet neuropsychologist Lev Vygotsky, that brains are culturally constituted and that the particular history, culture, and belonging of an individual to specific social institutions all participate in the way in which brains develop (Wolfe 2015). Neurological activity and synaptic communication are therefore influenced by a brain architecture whose constitution is embedded in broader social relations. This kind of knowledge may also challenge the neoliberal view of atomized individuals, all equal and therefore equally responsible for their conduct. Instead, individuals should be seen as the product of a particular combination of factors that would inform their behavior beyond their ability to choose otherwise.

In addition, studies reported by Sapolsky (2004) show that if brain damage in the prefrontal cortex occurs after the end of brain development in early adulthood, individuals are likely to display little capacity for foresight or projecting the consequences of their actions in the future in particularly emotional circumstances but the rest of their cognitive abilities will function fine. By contrast, when damage occurs at an earlier stage, individuals are likely to display premeditated antisocial behavior, a condition labeled as "acquired sociopathy" (Sapolsky 2004, p. 1793). It seems fair to then legally ask the question of the responsibility of these individuals whose biology has been altered at a stage which resulted in their current socially abnormal behavior.

This knowledge may lead the law to embrace even further the category of diminished responsibility (Hogson 2000) that is already present in terms of assessing whether or not the individual was responsible for his or her actions. The responsibility of the individual is diminished because neurolaw established the relevance and influence of factors external to the individual that affected his brain and the law then needs to create a new category of the legal subject. The neoliberal self is once again potentially challenged by such a use of neuroscience by the law.

In adulthood, neurolaw may also establish that the individual cannot be held responsible for certain biological modifications. The American case of Jeffrey Weinstein (Rosen 2007) and more recently the story of Terrence Martin (Johnston 2011) illustrate this point. Martin was diagnosed with Parkinson disease and further neuroscientific evidence revealed that he had become a sex addict as a result of the medication he was prescribed. In other words, his brain had changed and although he knew right or wrong, he was unable to act on this knowledge because his brain was driving him to behave in an overly sexually active manner.

As showed by Sapolsky, an individual can know the difference between right and wrong but be unable to act on this knowledge because of neurological malfunction (Sapolsky 2004). This suggests that brain plasticity—a notion to which this chapter will come back later—may actually show that the individual cannot be held fully responsible for changes affecting his brain and therefore his behavior

This point is explored further through a slightly different angle in this book through Melissa Parry's chapter on Spectral Personas which develops another issue potentially interfering with traditional reasoning around notions of responsibility associated with legal personhood. She mentions how brain implants and brain gene transfers for therapeutic purposes may have an impact on personality and personhood (Meloni et al. 2011) as they get somehow "absorbed" by the individual's brain and create unique neurological connections that cannot be differentiated anymore from the individual biological existence. Studies have shown that these biotechnologies, although often successful in treating the neurological conditions they were used for, have a number of side effects for individuals such as impulsivity, inability to make thoughtful choices or lack of regards for the consequences of their acts (Meloni et al. 2011). These developments matter because they undermine notions of legal responsibility while at the same time obscuring the differences between technologically driven and purely biological behavior. It becomes then difficult to think of the individual as autonomous in their choices and to require them to take responsibility for their actions.

As developed earlier, responsibilization is a key aspect of the neoliberal discourse and practice. It can be described as requesting the individual, through the law, moral diktats or other social conventions, to see himself as a free and responsible subject in terms of both the actions he engages in and their outcomes (Biebricher 2011). Neurolaw may however assist the individual to resist this responsibilization. Scientific facts and legal processes may be seen as coproducing a different kind of individual whose responsibility is not just diminished but who has been deemed irresponsible although not insane. Biebricher argues that with neuroscience, the individual, seeing himself as a "neurochemical" or "somatic" self (Rose 2007, p. 188) and through appropriate legal processes can be, in fact, engaging in a process of self-irresponsibilization (Biebricher 2011). This means that the individual is denying being in control of his biology and therefore cannot be held responsible for it.

Biebricher uses Foucault's concept of "counter-conduct in the sense of a struggle against processes implemented for conducting others" (Foucault 2008, pp. 200–201) to show how some forms of resistance can be identified in various contexts, that would include the legal context. When an individual defines himself primarily as suffering from neurological malfunction or chemical unbalance, it alleviates the burden of personal responsibility (Biebricher 2011) and therefore weakens the neoliberal discourse of responsibilization. Through this lens, criminal behavior is not the outcome of human choice but biological factors resulting in an individual being poorly equipped in decision-making skills. This representation of the self however carries other characteristics that may add complexity to this view.

The Return of Neoliberalism

This final section revisits some of the arguments developed previously and shows that at a deeper level, neurolaw is actually engaged in a process of consolidation of the ethos of neoliberalism.

Neoliberalism can be thought of as doing two things that may appear antithetical: it values freedom and liberty as paramount and advocates for liberating the individual from the rules and regulations imposed by state while at the same time, it leads to further regulations (Braithwaite 2008), more control and more surveillance of the individual both from institutional sources but more importantly from the individual himself self-regulating his behavior. Neoliberalism creates a representation of the individual who is constantly expected to gain knowledge about himself to increase his human capital and apply this knowledge to develop new capabilities, be more productive, more successful, and better adjusted to the political economy of late capitalism.

Neuroscience, through brain imaging, in particular gives us an insight into the way in which certain information is processed or triggers specific reactions. It can then be inferred that such a capacity to observe the functioning of one's brain can actually be a tool against the determinism that critiques often oppose to neuroscience. In fact, studies have shown that neurofeedback—real-time displays of brain activity—can play a positive role in self-regulation of the brain by the individual (Wasserman and Johnston 2014). In other words, seeing how our brain works may help us regulate its activity. If on the one hand this perspective restores to an extent our sense of agency, on the other, it brings responsibility back onto the individual to observe and know their brain in order to act on it.

Neurolaw could also be a tool to bring more responsibility back on the individual. An accused may be represented by the law as potentially irresponsible of committing the criminal act because it can be shown that his brain is abnormal, but at the same time deemed responsible for not attempting to know more about his biology and for not doing anything about it. In other words, the individual becomes expected to maximize all the information about his biology and not doing so may generate new legal consequences in terms of defining the level and kind of responsibility of the individual.

Seen as a dynamic mode of representation of human behavior, neoliberalism manages to displace responsibility when new scientific evidence shows that individual ability to act freely is somehow limited due to neuronal factors. This weakens the case for individual responsibility understood in a traditional way as the exercise of rational choices and the acceptance of the consequences of these choices. Neoliberalism finds new avenues to conceptually bring responsibility back in through focusing on aspects of human behavior that could counteract these biological limitations and as a result, it re-responsibilizes the individual by emphasizing his or her duty to overcome the limitations revealed by neuroscience.

As mentioned earlier, studies have shown that the brain is adaptable; it can change, to an extent, both its structure and the activity taken place within it. Neuroplasticity refers to this phenomenon and shows that the brain has the ability to modify itself based on experiences but also to repair and reprogram itself after an injury for instance (Green and Bavelier 2008). This view of the brain fits nicely into the neoliberal mantra of flexible individuals who can adapt and make new choices depending on new information. Neurolaw is certainly interested in this aspect of the science of the brain as it has consequences in terms of the level of responsibility of the individual who knew that something was wrong but did not address it.

Brain plasticity is also a key concept as it means that individuals can act on their biology, even at the brain level. Neoliberalism also expresses that the individual needs to overcome his or her biological limits through technology (Read 2009) in order to truly become an entrepreneur of himself. This representation of the human capacity to change oneself through modifying the structure and activity of the brain continues to represent the individual as a "neuronal" or "neurochemical" subject (Pitts-Taylor 2010, p. 639). A subject who can and must "manage" his brain through learning, mental training and other biological activities that will enhance his brain performances while at the same time avoiding risks that could lead to mental underperformance, loss of memory, and mental sicknesses of various sorts. Similarly to Wasserman, Pitts-Taylor (2010, p. 639) argues that this perspective can portray the individual as resisting determinism but she critically questions this assumption and instead proposes that it "...firmly situates the subject in a normative, neoliberal ethic of personal self-care and responsibility linked to modifying the body." In fact, following Ericson and colleagues (2000), it can be argued that this ethics of personal self-care replaces an ethic of state care and focuses on individual responsibility in ways highly compatible with the principles of market fundamentalism cherished by neoliberalism.

Biebricher analysis of self-irresponsibilization mentioned earlier also includes a critical conclusion about the fact that neuroscience and neurolaw ultimately consolidate a neoliberal representation of the individual who must accept to carry the responsibility for all his actions. In fact, with the knowledge available through neuroscience, including brain plasticity and neurofeedbacks, it becomes difficult to claim that a given behavior is solely the result of brain malfunction. In other words, people who may commit crimes but also people victim of difficult circumstances such as long-term unemployment, chronic depression, or repeated domestic violence situations may be considered responsible because they could have used the knowledge available to make better use of their brains. In this sense, the fact that they did not act on their brain make them accountable for their personal failure which leads to the conclusion that "neuroirresponsibilization is a form of counter-conduct under neoliberalism, once the individual has become neuro-scientifically literate, he returns to the neoliberal fold." (Biebricher 2011, p. 482).

Legal and criminology scholars have noticed that increasingly the law considers the individual who committed a criminal act and not solely the act in itself. Neurolaw may result in a radical reformulation of these notions if as detailed by Hogson (2000), we reframe the image of a criminal as a person with a particular sickness and therefore someone who needs treatment more than punishment. The act committed may be of a criminal nature, but the individual cannot be responsible because his brain displays abnormalities.

This approach may appear more humane and progressive than the current system and can be linked to the notion of a somatic self-developed briefly by Biebricher (2011) and who will use irresponsibilization as a strategy to avoid legal and moral responsibility. However, the cost of such a strategy can be extremely high. Mentally sick people require time in isolation from the rest of society. Criminals framed as mentally sick patients may be locked up for an indefinite period of time because the way their brain is wired represents a threat for society that cannot be mitigated (Wasserman and Johnston 2014). This could greatly increase the number of people deprived of their liberty and with no endpoint leading to their rehabilitation. This would ultimately increase the controlling and regulating power of government, a discrete but real and significant side effect of neoliberalism.

When chemical treatment for cognitive disorders becomes available and mandatory for criminals once they are framed as sick people, Wolfe draws our attention to the effect of this neurolegal knowledge on the general population (Wolfe 2015). It could be argued that drugs enabling criminals to control their emotions, to stay focus, to develop empathy for instance, should be made available to society to enable individuals to enhance their cognitive abilities and therefore participate in the neoliberal self's constant quest for biological improvement.

If Neoliberalism is built on notions of individual freedom and liberty, it also reinforced the power of regulatory institutions and ultimately enables governments and private actors to engage in vast operations of control and surveillance (Millei and Joronen 2016). In the criminal system, treating criminal as sick people opens up new avenues to establish monitoring processes in jails or other, new institutions with databases of personal biological data, detailed genetic make-up, brain structure and activity, or hormonal levels. This information will not be available to the individual but can be consulted by authorities when they consider it adequate and potentially used when necessary in a legal context. Pushing it one step further and neurolaw is in fact creating a new panopticon based on the medical surveillance of individuals who do not know when and how this surveillance occurs or to which effect.

Finally, the importance of the early years of a child has been well documented in terms of brain development (Tierney and Nelson 2009). The brain may not develop or function appropriately when a child grows in an insecure environment, with no or little care, affection, love, and attention. This can in turn explains certain criminal behavior due to an atrophied prefrontal cortex (Moir and Jessel 1995). Parents unable to provide such an environment may ultimately be held responsible for the criminal behavior of their adult child. Neoliberalism and neurolaw create therefore new representations of individual responsibility, more remote from the act. If one section of the population is "excused" through neuroscientific evidence, neoliberalism through neurolaw can recreate new sites of responsibilization and impose new regimes of (parental) responsibility. This confirms the conceptual ability of neoliberalism to displace responsibility in order to target new individuals that may bear the consequences of acts performed by other individuals whose responsibility cannot be established due to neurological particular conditions.

Conclusion

Neurolaw represents a propitious field to establish new forms of knowledge about the human experience as it benefits from the credibility of both scientific and legal established truths. At the same time, it is surreptitiously promoting a particular representation of the human experience that is limited and simplistic but that fits comfortably with a dominating neoliberal understanding of the self. In fact, legal personhood is evolving to accommodate the outcome of this process of knowledge co-creation that may result in a renewed responsibilization of the legal subject. It does so by accepting that individuals may not always be accountable for their actions due for example to brain malfunctions, but that they can and should be responsible for not acting on this information once they are cognizant of it.

Unlike the reductionist view presented by neurolaw and neoliberalism, human behavior is complex, at times paradoxical, and always multidimensional. It is the way in which biological, sociological, and psychological factors interacts that best explain the cause for human behavior (Schehr and French 2012). Ultimately, the representation of a legal subject must attempt to provide a holistic account of the human experience and stay away from a neurocentrism that cannot give a fair account of human nature.

Bibliography

Barry, A, Born, G, & Weszkalnys, G 2008, 'Logics of interdisciplinarity', *Economy and Society*, vol. 37, no. 1, pp. 20–49.

Becker, G 1964, *Human capital: a theoretical and empirical analysis with special reference to education*, Columbia University Press, New York.

Biebricher, T 2011, '(Ir-)Responsibilization, genetics and neuroscience', *European Journal of Social Theory*, vol. 14, no. 4, pp. 469–488.

Braithwaite, J 2008, *Regulatory capitalism: how it works, ideas for making it better*, Edward Elgar, Cheltenham.

Csuri, P & Mantilla, J 2015, 'Are we heading towards a brain-centered culture? An interview to Fernando Vidal', *Revistas Culturas Psi/Psy Cultures*, vol. 5, pp. 5–12.

Dumit, J 2004, *Picturing personhood: brain scans and biomedical identity*, Princeton University Press, Princeton.

Ericson, R, Barry, D & Doyle, A 2000, 'The moral hazards of neo-liberalism: lessons from the private insurance industry', *Economy and Society*, vol. 29, no. 4, pp. 532–558.

Foucault, M 2008, *The birth of biopolitics: lectures at the Collège de France, 1978–1979*, trans. G Burchell, Palgrave Macmillan, New York.

Green, CS & Bavelier, D 2008, 'Exercising your brain: A review of human brain plasticity and training-induced learning', *Psychology and Aging*, vol. 23, no. 4, pp. 692–701.

Harman, C 2008, 'Theorising neoliberalism', *International Socialism*, vol. 117, pp. 25–49.

Harvey, D 2005, *A brief history of neoliberalism*, Oxford University Press, Oxford.

Hogson, D 2000, 'Guilty mind or guilty brain? Criminal responsibility in the age of neuroscience', *The Australian Law Journal*, vol. 74, pp. 661–680.

Jasanoff, S (ed.) 2004, *States of knowledge: the co-production of science and the social order*, Routledge, London.

Jasanoff, S 2007, 'Making order: law and science in action', in EJ Hackett et al. (eds), *The handbook of science and technology studies*, 3rd edn, MIT Press, Cambridge.

Johnston L 2011, 'Child sex charges: former MP set free as judge deems Parkinson's drugs to have caused hypersexuality', *Sydney Morning Herald*, 30 November.

May, C 2009, *The global political economy of intellectual property rights*, 2nd edn, Routledge, London.

Meloni, R, Mallet J & Faucon Biguet, N 2011, 'Brain gene transfer and brain implants', *Studies in Ethics, Law, and Technology*, vol. 4, no. 3, pp. 1–17.

Millei, Z & Joronen M 2016, 'The (bio)politicization of neuroscience in Australia early years policies: fostering brain-resources as capital', *Journal of Education Policy*, vol. 31, no. 4, pp. 389–404.

Moir, A & Jessel, D 1995, *A mind to crime: The controversial link between the mind and criminal behaviour*. Michael Joseph Ltd.

Parry, B 2018, 'The social life of "scaffolds": examining human rights in regenerative medicine', *Science, Technology, & Human Values*, vol. 43, no. 1, pp. 95–120.

Pickersgill, M 2013, 'The social life of the brain: neuroscience in society', *Current Sociology*, vol. 61, no. 3, pp. 322–340.

Pitts-Taylor, V 2010, 'The plastic brain: neoliberalism and the neuronal self', *Health*, vol. 14, no. 6, pp. 635–652.

Pustilnik, A 2008, 'Violence on the brain: a critique of neuroscience in criminal law', *Harvard Law School Faculty Scholarship Series*, available at http://lsr.nellco.org/harvard_faculty/14.

Read, J 2009, 'A genealogy of homo-economicus: neoliberalism and the production of subjectivity', *Foucault Studies*, vol. 6, pp. 25–36.

Rose, N 2007, *The politics of life itself: biomedicine, power, and subjectivity in the twenty-first century*, Princeton University Press, Princeton.

Rosen J 2007, 'The brain on the stand', *New York Times Magazine*, 11 March.

Sapolsky, R 2004, 'The frontal cortex and the criminal justice system', *Philosophical Transactions of the Royal Society of London. Series B: Biological Sciences*, vol. 359, pp. 1787–1796.

Schehr R & French C 2012, 'Mental competency, law and plea-bargaining: a neurophenomenological critique', *Albany Law Review*, vol. 79, no. 3, pp. 1091–1182.

Schultz, T 1972, 'Human capital: policy issues and research opportunities', *Human Resources*, vol. 6, pp. 1–84.

Slaby, J 2010, 'Steps towards a critical neuroscience', *Phenomenology and the Cognitive Sciences*, vol. 9, pp. 397–416.

Thatcher, M 1987, 'Interview for "woman's own"', *Margaret Thatcher foundation: speeches, interviews and other statements*, London.

Tierney, A & Nelson, C 2009, 'Brain development and the role of experience in the early years', *Zero Three*, vol. 30, no. 2, pp. 9–13.

Vidal, F 2009, 'Brainhood, anthropological figure of modernity', *History of the Human Sciences*, vol. 22, no. 1, pp. 5–36.

Wasserman, D & Johnston, J 2014, 'Seeing responsibility: can neuroimaging teach us anything about moral and legal responsibility?', *Interpreting Neuroimages: An Introduction to the Technology and Its Limits*', Special Report, Hastings Center Report, vol. 45, no. 2, pp. 37–49.

Wolfe, C 2015, 'Brain theory between utopia and dystopia: neuronormativity meets the social brain', in M Pasquinelli (ed.), *Alleys of your mind: augmented intelligence and its traumas*, Meson Press, Lüneburg, pp. 173–184.

Zamora, D 2014, 'Foucault's responsibility', *Jacobin*, 15 December, https://www.jacobinmag.com/2014/12/michel-foucault-responsibility-socialist/.

CHAPTER 4

Legal Personhood in Postgenomic Times: Plasticity, Rights, and Relationality

Marc de Leeuw and Sonja van Wichelen

Introduction

Legal personhood performs an important function in modern law and legal theorists indicate that legal persons can be approached from a *legalist* and *realist* perspective (see Naffine 2003). While the first view attends to persons as masks, that allow legal relationships to solve certain problems arising within the framework and rules of law, the latter refers to the natural person, and seeks to stabilize the person by equating it as much as possible to a biological and metaphysical "given" of the human. Prompted by a variety of contemporary issues—including feminist, postcolonial, environmental, and posthuman concerns—several scholars in jurisprudence argue for a return to the Roman model

M. de Leeuw (✉) · S. van Wichelen
Sydney, NSW, Australia
e-mail: m.deleeuw@unsw.edu.au

S. van Wichelen
e-mail: sonja.vanwichelen@sydney.edu.au

M. de Leeuw and S. van Wichelen (eds.),
Personhood in the Age of Biolegality, Biolegalities,
https://doi.org/10.1007/978-3-030-27848-9_4

of personhood (Grear 2013; Mussawir and Parsley 2017). Rather than representing a naturalized person, such a model avoids any reference to something natural, living, or organic. Instead, the model can take up any entity and can be shaped by deliberate techniques. In her insightful article "Laws Entities," Anna Grear (2013) points to this model's plasticity: It is infinitely malleable, and since it is not locked to any pre-given construction, it can openly respond to complexity.

While conceptualizations of a plastic personhood are taking place in jurisprudence, the idea of a biological plasticity is emerging in social studies of the new biosciences. Bodies, brains, genes, and cells are conventionally thought of as bounded entities of life. Yet, developments in the new life sciences—such as microbiomics, neuroscience, epigenetics, and immunology—are challenging this perception and describe biological life instead as plastic. Plasticity refers to flexibility and openness of the body and its internal functions to the body's environment. Here, biological processes are less linear and determinant as the sciences once thought. In its place, bioscientific insights point to complex permeabilities and malleabilities.

Our chapter should be regarded as a preliminary thinkpiece. One that examines these two conceptual approaches to plasticity. What does plasticity mean in the study of law's persons? And what does it mean in the study of postgenomic science? We describe how these conceptualizations carry with them a clear social and political project—one that we support and underscore—but that taken together run into some kind of trouble. This trouble becomes evident when turning to existing and future claims of biolegality from people and communities that are based on biological rights. These rights claims, we argue, go against plasticity in both law and science. It raises the question of how publics and states can lay claim to artificial forms of legal personhood, devoid of any reference to the natural person, in a time marked by biolegitimacy.

Troubled Persons

A renewed interest is growing in the study of legal personhood. We witness this in relation to developments in bioscience and biomedical technologies (Pottage 2002; Van Beers 2017; Selkälä and Rajavuori 2017; De Leeuw and Van Wichelen, forthcoming; Wishart 2016), in discussions around Artificial Intelligence (Lawrence and Brazier 2018; Teubner 2006), and in debates surrounding animal, environmental, and earth rights (Berg 2007; Cohen and Regan 2001; Grear 2010, 2013; Sunstein and Nussbaum 2004; Vatter and De Leeuw 2019). Taken together, the disposition of this growing literature is to describe recent developments in science and society that trouble legal personhood, to situate legal personality in a historical and anthropological context, and to propose new ways that allow law's persons to accommodate and integrate the complexities of contemporary science and society.

The trouble that contemporary science and society brings to legal personhood has much to do with modern law's distinction between the artificial and the natural person, a distinction used to differentiate between human bearers of rights on the one hand, and juridical persons such as corporations on the other. The stabilization of the natural person occurred over an extended period, starting from the influence of Christianity on law, to the rise of the human and life sciences in the nineteenth century. Biology played a particularly vital role in constituting the naturalized person. Apart from the rudimentary facts of biology that have organized law foundationally in the realms of family and inheritance—such as birth, death, sex, and reproduction—the biological and psychological sciences helped constitute the natural person by providing scientific knowledge of humanity and human behavior. The strong influence of human rights in legal systems around the world further emphasized the legal subject's reference as a naturalized human being. It was not until the start and impact of the Human Genome Project that the biosciences and their technologies transformed legal practice, from applications in forensics and (bio)security, to foundational changes in family law.

The natural person in modern law tightly references the natural body; through birth and death, the division of the sexes and metaphysical human (and liberal) traits such as autonomy, integrity, and agency. This bounded human self received due critique from feminist scholars, who pointed out discrepancies of the natural person when turning to the pregnant body (Karpin 1992; Davies and Naffine 2001; Nedelsky 1990). Anthropological theory too subjected law's natural person to reproach by pointing to the Eurocentric constructions of personhood, including metaphysical understandings of autonomy and bodily integrity (Collier et al. 1995; Comaroff and Comaroff 2001; Strathern 1988). Rather than the bounded self, individuality in non-western societies is relational, or as Marilyn Strathern calls it "dividual" (1988, p. 13). Likewise, compared to modern societies, studies of primitive and premodern societies indicate a different relationship between persons and things, the social and the economic, kinship and markets (Kopytoff 1986; Mauss 1985; Strathern 1999), which disturb the foundational modern distinction between a person and a thing. More recently, legal anthropologists have transplanted this discussion in the realm of law by bringing into perspective complex reconfigurations of personality and property through emerging biotechnologies (Pottage 2004, 2007; Strathern 2005).

The main tenet of contemporary discussions in law and biotechnology is that they blur the distinction between artificial and natural personhood. Biological traits that are traditionally attached to natural persons are no longer certain. As Van Beers describes, natural persons now extend the human species (for example, xenotransplantation, cybrids, extending personhood to animals), the beginning of natural personality does not necessarily coincide with birth (for example, reproductive technologies allow for fetal personhood), the ending of personality does not necessarily coincide with death (for example, brain death, cryopreservation), natural persons are not only the product between persons of the opposite sex (recombinant and blended families), and finally, natural persons themselves in fact extend the category of male and female (transgender, third sex) (2017, pp. 563–566). These developments change and fragment natural personhood. Van Beers cautions that a total fragmentation can lead to the legal system becoming too singular, and therefore too weak to guarantee certainty and equality for the law (ibid., p. 591).

Within these developments the concept of plasticity has taken hold in both law and science. On the one hand, legal plasticity is proposed to respond more adequately to the changes in science and society, on the other, biological plasticity is offered as a conceptual explanation in understanding the new postgenomic body. How—if at all—can these conceptualizations assist one another in future problems of personhood and the new biosciences?

Legal Plasticity in the Age of Global Capital

The distinction between artificial and natural personhood was not always in place. In fact, the birth of legal personhood during the Roman epoch did not engage with the modern distinction between a person and a thing. Instead, personhood fulfilled a clear fictionalized function to solve a problem in the realm of institutions and operated primarily as a juridical placeholder to accommodate relationships with for instance the city, estates, or kin (Mussawir and Parsley 2017, p. 49). Moreover, legal personhood did not embody any reference to natural or human personhood; it was not shaped according to biological understandings of human beings. As Alain Pottage indicates, Roman law "acknowledged that the biological or psychological substrate (the role player) was tailored to the specifications of the role, which meant that the law fictionalized biology so as to provide the necessary warrant for each of the transactional personae that it invented" (2002, pp. 289–290). The example of Roman adoption (see Mussawir and Parsley 2017) is an apt description of this operation. Adoption was not—as it is now—a legal technology that operates on embodying "as if" blood relationships (see Nash 2005; Yngvesson 2002). Instead, the biological order could be manipulated to accommodate certain juridical transactions; a grandson could be adopted to the level of son, or a son could be readopted to the level of father of another son, and so on.

Several legal theorists argue that the technicalities of this manipulation, fashioning, or "machining" of persons offer great potential and capacity to reconfigure the person in our times. Drawing on and extending the work of Ngaire Naffine, Anna Grear (2013) proposes a general theory of legal subjectivity that empties the person of anything natural or biological. Instead, the ideal legal person should be plastic, meaning that the concept should have a systematic and exhaustive capacity to respond flexibly and malleably to complexity. Away from the natural person, Grear's person is a legal entity that has the "formal capacity to bear a legal right and so to participate in legal relations" (ibid., p. 85). In its plasticity, and therefore limitless formation, it functions as a placeholder making explicit the limits of law and the gap between the legal person and the biology or nature of the entity itself.

There's a clear social and political project at the heart of the plastic model of legal personhood (or entity-hood as Grear would call it). As Grear states: "In an age of mounting pressure concerning the endangerment of animals, ecosystems and so forth, [entity-hood] possesses theoretically flexible, virtually unlimited potential for multiplying law's entities while usefully and *simultaneously*, pointing to law's strategies and limits" (ibid., p. 86). The model would de-anthropomorphize personhood, extending rights to nonhuman animals and environments in the natural world but also to entities in the artificial world such as robots and other forms of Artificial Intelligence (AI). Here, entities become the subject of rights rather than "rights bearers," a major shift from the human rights frameworks that have informed these issues in the past. The plasticity of the model allows for openness and inclusiveness at a time when legal personhood is dominated by global capital, one where corporate personhood is profiting more from naturalized forms of legal personality than any human being is able to.

Away from the social project, entity-hood allows a return to the technicalities of Roman jurisprudence, in which personhood can be fashioned and machined. While it promises great potential by offering an inventive and creative mode of technical precision, returning—as Mussawir and Parsley maintain—to the juridical art of the person (2017, pp. 50–52), we argue in this chapter that it is not without its problems. In its first instance, entity-hood does not embody or link with any sociopolitical context. It is only in the materialization of legal entity-hood that we can observe and analyze how culture, knowledge, and power are affecting its construction. Ironically, the proposal of entity-hood—empty of any situational context—comes at a time when biology is starting to refine environments at a molecular level. The next section describes a few of these developments and brings these changes to bear on what has been said about legal plasticity.

Biological Plasticity and the Postgenomic Body

The life sciences today are profoundly delivering and shaping new understandings of human life. This development is due to emerging bioscientific developments phrased as the postgenomic age, a time in which permeable considerations of the body replace gene-centric explanations of health and disease. In contrast to the more deterministic view of biological life as relatively stabilized and bounded, postgenomic explanations of life render cells, genes, brains, and bodies as plastic—meaning, as much more flexible, open, porous, and subjected to the vagaries of the environment.

Scholars in the social study of science have analyzed these forms of plasticity in the field of neuroscience (Rose and Abi-Rached 2013; Pitts-Taylor 2016; Rees 2016), epigenetics (Landecker 2007; Meloni 2016, 2019; Mansfield and Guthman 2015; Mansfield 2017), microbiomics (Lorimer 2017; Schrader 2017), immunology (Hird 2007; Fannin 2014; Martin 2010), and more generally, in the field of bioscience and social theory (Rose 2007, 2013; Mansfield 2017; Richardson and Stevens

2015; Meloni 2019). Notwithstanding the different variations and implications of plasticity in the developments of these different biosciences, they all have in common that plasticity contributes to a more complex understanding of embodiment, one that shifts or softens the boundaries of the human body, of body parts, cells, and genes. Hence, epigenetics for instance, explains how stress, toxins, and food can impact the structure and action of genes; studies in neuroplasticity show how the adult brain proves generative of new neuronal circuits; research on the microbiome describes how the gut system is dependent on microbes in our direct environment; and scholarship on microchimerism in immunology show that women retain fetal cells from children she gestated. By modeling these developments around the conceptualization of plasticity, social theorists emphasize the indeterminate and unbounded nature of the body and its biological processes.

Like the uptake of legal plasticity in legal theory, the study of biological plasticity in social theory has a clear social and political project. Plasticity functions as an antidote to the "biology of fixedness," a paradigm that dominated the Western life sciences for the past centuries, and that favors demarcated binaries between inner and outer body, nature, and culture (Meloni 2018, p. 3). These demarcations have been responsible for essentializing differences in sex and race, fueling ideologies in the realm of politics and governance. With biological plasticity, such demarcations are presumably no longer viable, or they are much harder to sustain. Feminist philosophers in new materialism have particularly emphasized the potential that biological plasticity can have for feminist theory and practice (Barad 2012; Hekman 2010; Kirby 2011).

Of course, the concept of plasticity—and its inherent project of combating biological fixedness—is not new. During the height of postmodern theory in the 1980s and 1990s, plasticity functioned as the paradigmatic figuration that offered bodies a way out of biologized regimes of naturalness and discussions of plastic surgery rendered the plastic body as the "political object par excellence" (Bordo 1993, p. 245; Grosz 1994). But as we learned that these postmodern notions of plasticity were also capable of reinforcing some of the same disciplinary regimes it sought to dismantle, so too have social theorists of science more recently cautioned against the progressive appraisals of biological plasticity.

Some scholars base their caution through their empirical research, which reveals that the science of biological plasticity as practiced today does not presuppose the progressive idea of the boundless self, the fluidity of sex, or the irrelevance of race and ethnicity (Richardson 2017; Lock 2015). Others illustrate how the sociopolitical ramifications of biological plasticity can be as much oppressing to populations as the mechanism found in the biology of fixedness (Landecker and Panofsky 2013; Van Wichelen 2016). Epigenetics has in particular been met with social and political concern in that it has been used to renew and reproduce deterministic views of disease alongside markers of class and race (Mansfield 2012; Guthman and Mansfield 2015). As Sarah Richardson argues, "discourses of epigenetic plasticity encourage individuals to see themselves as protectors and cultivators of their own epigenetic health, inviting neoliberal discourses of health optimization as individual responsibility" (2017, p. 44). This caution invites us to think anew about the concept of plasticity in legal theory. Which frictions occur when biological plasticity and their social reiterations meet legal plasticity in the realm of rights and juridical practice?

Rights Claims Informed by the New Biologies

An important determinant in new arrangements of persons and things is the subject of capital (Cooper 2008, p. 19; see also Rajan 2006). While the emergence of bioeconomies includes actual industries such as pharmaceutical or biotech companies, they also pertain to the more intricate markets in the field of reproduction (assisted reproductive technologies, surrogacy, and egg banking) or genetic testing. Scholars in sociology and anthropology have argued for the productive relationship between bioeconomies and (moral, civic or legal) subjectivity, pointing to the composition of biosociality (Rabinow 2005), biological citizenship (Petryna 2013), biolegitimacy (Fassin 2009), and biolegality (Lynch and McNally 2009; De Leeuw and Van Wichelen, forthcoming). This scholarship—we argue—illuminates two developments in society and law. One is that knowledge pertained from the biosciences are reshaping new understandings of self, property, kin, and community. And that these new understandings are co-constitutive of new legalities and legal forms.

Returning to the question of legal personhood, our main problematic here is to see how postgenomic knowledge is (a) translated to the domains of naturalness and artificiality, and (b) what impact potential legal entity-hood would have on future claims and rights that are based on postgenomic knowledge. This is a big question, and one that is too early to answer meaningfully because postgenomic claims are still scarce in the legal realm. Nonetheless, appropriations of postgenomic science have been surfacing quite strongly in the public sphere and popular culture (see Richardson 2017). In particular, some claims and justifications for certain rights and entitlements have been made in the realm of reproduction and involving commercial and cross-border surrogacy (Payne 2016; Van Wichelen 2016) and in the realm of indigenous rights (Warin et al. 2018).

In the case of reproduction, epigenetics positions the pregnant woman as transgenerational vector of harm. Stress, toxins, and food can alter the genetic expression of the child and her respective generations. The science suggests that they can be socially, legally, and medically responsible for, not only her "future child," but also for the future generations of her offspring (see for instance Van Wichelen 2016). While this could open the door for allegations of "wrongful life" and claims made on behalf of the "unborn child" or "future child," these categories are not legal persons in law; they cannot be categorized in the naturalistic viewpoint of the legal person. Advocates of a constructionist view of legal personhood—who propose conceptions such as entity-hood—argue that plasticity helps extend rights to the unborn or future child and will protect their interests.

In the case of indigenous rights, postgenomic science has been taken up in the field of indigenous politics and social justice, in which epigenetics is used as leverage in political discussions of reparation (see for instance Warin et al. 2018). The science of environmental epigenetics provides indigenous communities a context in which to connect the colonial and traumatic past to their biological existences today. Within the legal field, such claims have the likelihood of broadening the context of "Discovery of Harm" to include multigenerational and intergenerational effects of historical trauma.

In surrogacy, postgenomic knowledge brings a new perspective to the rights and responsibilities of the pregnant surrogate mother. Since the *Johnson versus Calvert* case in 1993—where motherhood was assigned to the genetic parent in a Californian commercial surrogacy arrangement, dismissing any entitlement to parenthood of the gestational surrogate—genetic motherhood was effectively distinguished from gestational motherhood and bypassed the unique legal principle of *mater semper certa est* (the mother is always certain). This landmark case led to the dominant tendency in courts today that the genetic parent's rights surpass the rights of persons who gestated their child in surrogacy arrangements.

Postgenomic knowledge (like epigenetics, microbiomics, and chimerism), however, elicit a reconfiguration of relationality with the fetus. Not only does the science give evidence that the surrogate provides important biological contributions to the fetus by way of food intake and the make-up of her microbiome, her environment can have a significant impact on the development of the child by way of stress and environmental toxicity. Furthermore, as Margrit Shildrick (2019) argues, fetal microchimerism allows the biological entganglement of genetic parents with both the gestational mother and fetus.

These developments have important consequences for the bioethics and legal regulation around gestational surrogacy. While *Johnson versus Calvert* based their case on the assumption that no biological relatedness exists between surrogate and child, postgenomic science challenges these assumptions. Suggestions on how to address these changes in current policies and regulations around surrogacy have been made (Loike and Fischbach 2013) and point to science communication as well as definitional changes to informed consent. Moreover, feminists use the science to voice their stance against surrogacy (Hudson 2018) and form new legitimacies that give gestational surrogates the same rights to parentage as traditional surrogates who use their own eggs in surrogacy arrangements.

Important for our point here is that these developments allow for a manifestation of "rights" that is informed by biology and sustained by references to science. Regarding legal personhood, such "rights" translate a biological relationality into the category of natural personhood, thereby naturalizing it even more. The problems associated with natural personhood, detailed by many legal scholars, find a new manifestation in the incorporation of postgenomic evidence. While genomic evidence is much critiqued for producing new forms of discrimination (Dolgin 2000; Duster 2015; Lemke 2005, 2013), there is no reason to think that this could not trigger a similar effect with postgenomic evidence.

In combating the further naturalization of personhood, "entity-hood" is offered as a way out of the selective particularities of natural personhood that have contributed to discriminatory and exclusionary categories. However, we are not entirely convinced that "entity-hood," which radical potential lies in its "emptiness," can stand for biological life that is characterized by its "fullness." What we mean by "fullness" does not only relate to the postgenomic body, but to the more philosophical idea of *bios* as qualified life. The constructionist perspective on legal personhood still relies on the framework of rights and interests (shifting the focus to the object of rights rather than rights bearers). But even when we turn to the two cases of "wrongful life" and "discovery of harm" we run into trouble, for rights and interests fail to articulate not only the extent to which environments are embedded in bodies, and our biological lives are entangled in inter- and multigenerational histories and biographies, but also the social, cultural, and political ways in which bodies are conceptualized and known.

Critical Notes to Plasticity

In coining the term "biolegality" Michael Lynch and Ruth McNally refer to changes in forensic practice in which DNA is constructed as a "machinery of truth." Apart from establishing a historical relationship between innovation and regulation, they argue that the exceptional reliance on DNA also assumes an epistemic relation in which "biological 'truth' justifies exceptional legal procedures" (2009, p. 296). Elsewhere, we advocate the use of biolegality as a heuristic term beyond the field of forensics and the dominant influence of genetics (De Leeuw and Van Wichelen, forthcoming). Instead, we argue that biolegalities emerge in and outside of legal frameworks, encompass all forms of biology, including the new biologies, and fulfill a vital role in the development of socio-legal theory. While biolegalities pushes legal theory to look anew at their premises in natural law, they provide socio-legal theory with novel empirical cases that trouble social and socio-legal understandings of nature and nurture, biology and culture, the grown and the made.

The biolegality of postgenomic personhood assumes biological plasticity. In contrast to legal plasticity—which relies on the "emptiness" of personhood—biological plasticity can only be conceptualized as "full." What we mean by this is that postgenomic knowledge defines the biology of humans as dependent on environmental influences. It is full in that the inclusionary politics lie not in its "emptiness" but rather in the capacity of biological life to process and metabolize "environment". Legal claims depend on a *biological* notion of relationality with this environment, and here the question arises how this notion of biological relationality can materialize in the conceptualization of legal plasticity. We propose in this chapter that the materialization of postgenomic claims to personhood (or entity-hood) cannot take place without a conceptualization of the body.

The scholarship of Marilyn Strathern points to potentialities contained in "the body," meaning that potentialities associated with the body give way to how certain claims (for instance ownership and rights) are made forceful in legal situations. In Western law, the oscillation between person and thing is a core mechanism through which we actualize the body. For instance, the potentiality of personhood for an embryo varies according to whether the embryo is frozen or thawed, implanted in utero or discarded (1999, p. 175). The body can actualize as a different "form" according to the relation that is actualized by the claim. Strathern notes that the question whether body parts can be seen as property is dependent not only on biological relations (the detachment of parts from the whole), but also *conceptual* relations (the conceptual capacities to detach parts from wholes). While biotechnology—such as reproductive technologies—enabled the detachment of body parts from their wholes, bioethics qualified the *conceptual process* of detachment, which continued the oscillation between persons and things via the form of the body. The emergence of postgenomic knowledge, however, further complicates the biology of parts and wholes as well as the conceptuality of detachment.

Entity-hood seems like a logical place to start when the emptiness of the legal concept can encompass the fullness of biological plasticity. But claims based on biological plasticity alone run the risk of further naturalizing personhood, albeit relational forms of personhood. Rather than facilitating a communitarian form of relationality, it emphasizes a biological form of relationality that reiterates Euro-American and modern assumptions of property. To put it differently, whereas claims on conceptual relationality are based on "dividuality," claims on biological relationality are based on an individualized biological plasticity. Ultimately, the claims and materialization of successful "entitihood" rely on the conceptual ways in which bodies are actualized. Hence, we suggest that legal plasticity can only be useful when it is able to account for the *conceptual* understanding of relationality that is actualized through the body. Such an engagement—we argue—is necessary if the intention of formulating new understandings of legal personhood is to combat biocentric, determinist, and binary understandings of nature and society.

The importance of the body in political theories of rights is also emphasized by the political philosopher Roberto Esposito (2015). He critiques the foundational distinction between persons and things in modern law and argues that it is part and parcel of what he terms "the immunization paradigm." The liberal subject of law is uniquely immunized from life. It conceives of itself firstly in terms of individuated persons, and only secondly as a social body. As Esposito argues, this immunitary logic can lead to the destruction of the body and the body politic by its negative form of protection and the negation of life. From this perspective, relegating the postgenomic body to the realm of either natural personhood or entity-hood reproduces the same dichotomy of understanding biological life either as person or thing. Since the body or the conceptual relation between person and thing cannot occupy a space in existing perspectives of legal personhood, our task is to articulate new avenues of thinking personhood beyond rights and interests.

Conclusion

Our chapter tentatively examined how the "new" plasticity of human nature—evoked by postgenomic research—assists or challenges the understanding of plasticity in law and legal personhood. While legal personhood evolved from a pure functional fiction (masks) in Roman law to encompassing a naturalized fiction (humans), its conceptual plasticity allowed us to apply this "nature" to corporations (as "incorporated persons") dissolving the actual biological body of persons into things, entities, and actual persons (meaning those who have rights). If the body can no longer be defined as containing or beholding a clear biological center—due to its microbial or epigenetic relationality—how can the law still attempt to individualize the body within concepts such as personhood?

There is great enthusiasm among legal scholars to reconfigure personhood, and the development of "entity-hood" is one clear proposal. Proponents argue that it allows inventive and creative modes of technical manipulation. Yet, as we argued, much depends on the materialization of legal entity-hood, and in our chapter, we emphasized the importance of conceptualizing relationality. Since bodies are central to actualizing relationality it is paramount to take the body into account. This is not the body as we have come to know it in modernity (one that is bounded by a rational self) nor for that matter, the one advocated by postmodernism (which is rhizomatic and fluid). Rather, for entity-hood to succeed we will need to attend to the body as actualized in an anthropological way. Law—as a site that privileges conceptuality—is the place *par excellence* to start this experiment.

Our attempt in this chapter should be regarded as a *diagnosis* of legal personhood amidst the conceptualization of plasticity in theorizing biological humanity. Our assessment is that plasticity in postgenomic science could potentially allow a move away from biological fixedness, but that plasticity in legal entity-hood runs the risk of fragmenting personhood in such a way that it becomes too singular to warrant legal certainty (see also Van Beers 2017). Rather than contributing to the plight of justice, our preliminary analysis suggests that plasticity—in both biology as well as law—can be as much part of an immunitary logic if legitimacies remain anchored in either modern scientism (where relationality can only be understood from a biological perspective), or in modern law (where relationality can only be understood from the liberal perspective of rights and interests). Rather than facilitating a communitarian form of relationality, the two approaches to plasticity emphasize either a biological or legal form of relationality that reiterates Euro-American and modern assumptions of property. To put it differently, whereas claims on conceptual relationality are based on "dividuality," claims on biological relationality are based on an individualized biological plasticity. Ultimately, the claim and materialization of justice relies on the conceptual ways in which bodies are actualized. We suggest that legal plasticity can only be useful when it is able to account for the *conceptual* understanding of relationality that is actualized through the body. Such an engagement is necessary if the intention of formulating new understandings of legal personhood is to combat biocentric, determinist, and binary understandings of nature and society.

References

Barad, K 2012 'Interview with Karen Barad', in R Dolphijn & I van der Tuin (eds), *New materialism: interviews & cartographies*, Open Humanities Press, Michigan.

Berg, J 2007, 'Elephants and embryos: a proposed framework for legal personhood, of', *Hastings Law Journal*, vol. 59, no. 2, pp. 369–406.

Bordo, S 1993, *Unbearable weight: feminism, western culture, and the body*, University of California Press, Berkeley.

Cohen, C & Regan, T 2001, *The animal rights debate*, Rowman & Littlefield, Lanham.

Collier, JF, Maurer, B & Suarez-Navaz, L 1995, 'Sanctioned identities: legal constructions of modern personhood', *Identities*, vol. 2, no. 1–2, pp. 1–27.

Comaroff, JL & Comaroff, J 2001, 'On personhood: an anthropological perspective from Africa', *Social Identities*, vol. 7, no. 2, pp. 267–283.

Cooper, M 2008, *Life as surplus: biotechnics and the transformations of capital*, University of Washington Press.

Davies, M & Naffine, N 2001, *Are persons property? Legal debates about property and personality*, Ashgate, London.

De Leeuw, M & Van Wichelen, S forthcoming, *Biolegalities: a critical intervention*, Palgrave Macmillan, London and New York.

Dolgin, JL 2000, 'Personhood, discrimination, and the new genetics', *Brooklyn Law Review*, vol. 66, no. 3, pp. 755–822.

Duster, T 2015, 'A post-genomic surprise: the molecular reinscription of race in science, law and medicine', *The British Journal of Sociology*, vol. 66, no. 1, pp. 1–27.

Esposito, R 2015 *Persons and things: from the body's point of view*, Wiley, New York.

Fannin, M 2014 'Placental relations', *Feminist Theory*, vol. 15, no. 3, pp. 289–306.

Fassin, D 2009, 'Another politics of life is possible', *Theory, culture & society*, vol. 26, no. 5, pp. 44–60.

Grear, A 2010, *Redirecting human rights: facing the challenge of corporate legal humanity*, Palgrave Macmillan.

Grear, A 2013, 'Law's entities: complexity, plasticity and justice', *Jurisprudence*, vol. 4, no. 1, pp. 76–101.

Grosz, E 1994, *Volatile bodies: toward a corporeal feminism*, Indiana University Press.

Grosz, E 2018, *Space, time and perversion: essays on the politics of bodies*, Routledge, London and New York.

Hekman, S 2010, *The material of knowledge: feminist disclosures*, Indiana University Press, Bloomington.

Hird, MJ 2007 'The corporeal generosity of maternity', *Body & Society*, vol. 13, no. 1, pp. 1–20.

Hudson, V 2018, 'Getting wiser on surrogacy', *Mercatornet*, 5 January, https://www.mercatornet.com/features/view/getting-wiser-on-surrogacy/20864.

Karpin, I 1992, 'Legislating the female body: reproductive technology and the reconstructed woman', *Columbia Journal of Gender and Law*, vol. 3, pp. 325–349.

Kirby, V 2011, *Quantum anthropologies*, Duke University Press, Durham, NC.

Kopytoff, I 1986, 'The cultural biography of things: commoditization as process', in Arjun Appadurai (ed.), *The social life of things: commodities in cultural perspective*, Cambridge University Press, Cambridge.

Landecker, H 2007, *Culturing life*, Harvard University Press, Cambridge.

Landecker, H & Panofsky, A 2013, 'From social structure to gene regulation, and back: a critical introduction to environmental epigenetics for sociology', *Annual Review of Sociology*, vol. 39, pp. 333–357.

Lawrence, D & Brazier, M 2018, 'Legally human? "Novel beings" and English law', *Medical Law Review*, vol. 26, no. 2, pp. 309–327.

Lemke, T 2005, 'Beyond genetic discrimination: problems and perspectives of a contested notion', *Genomics, Society and Policy*, vol. 1, no. 3, pp. 22–40.

Lemke, T 2013, *Perspectives on genetic discrimination*, Routledge, London and New York.

Lock, M 2015, 'Comprehending the body in the era of the epigenome', *Current Anthropology*, vol. 56, no. 2, pp. 151–163.

Loike, JD & Fischbach, RL 2013, 'New ethical horizons in gestational surrogacy', *Journal of Fertilization: In Vitro—IVF-Worldwide, Reproductive Medicine, Genetics & Stem Cell Biology*, vol. 1, no. 2, pp. 1–4.

Lorimer, J 2017, 'Parasites, ghosts and mutualists: a relational geography of microbes for global health', *Transactions of the Institute of British Geographers*, vol. 42, no. 4, pp. 544–558.

Lynch, M & McNally, R 2009, 'Forensic DNA databases and biolegality', in P Atkinson, P Glasner & M Lock (eds), *Handbook of genetics and society*, Routledge, London.

Mansfield, B 2012, 'Race and the new epigenetic biopolitics of environmental health', *BioSocieties*, vol. 7, no. 4, pp. 352–372.

Mansfield, B 2017, 'Folded futurity: epigenetic plasticity, temporality, and new thresholds of fetal life', *Science as Culture*, vol. 26, no. 3, pp. 355–379.

Mansfield, B & Guthman, J 2015, 'Epigenetic life: biological plasticity, abnormality, and new configurations of race and reproduction', *Cultural Geographies*, vol. 22, no. 1, pp. 3–20.

Martin, A 2010, 'Microchimerism in the mother(land): blurring the borders of body and nation', *Body and Society*, vol. 16, no. 3, pp. 23–50.

Mauss, M 1985, 'A category of the human mind: the notion of person; the notion of self', in M Carrithers, S Collins & S Lukes (eds), *The category of the person: anthropology, philosophy, history*, Cambridge University Press, Cambridge, pp. 1–25.

Meloni, M 2016, *Political biology: science and social values in human heredity from eugenics to epigenetics*, Palgrave Macmillan.

Meloni, M 2018, 'A postgenomic body: histories, genealogy, politics', *Body & Society*, available at http://journals.sagepub.com/doi/abs/10.1177/1357034X18785445.

Meloni, M 2019, *Impressionable biologies: from the archaeology of plasticity to the sociology of epigenetics*. Routledge, New York.

Mussawir, E & Parsley, C 2017, 'The law of persons today: at the margins of jurisprudence', *Law and Humanities*, vol. 11, no. 1, pp. 44–63.

Naffine, N 2003, 'Who are law's persons? From Cheshire cats to responsible subjects', *The Modern Law Review*, vol. 66, no. 3, pp. 346–367.

Nash, C 2005, 'Geographies of relatedness', *Transactions of the Institute of British Geographers*, vol. 30, no. 4, pp. 449–462.

Nedelsky, J 1990, 'Law, boundaries, and the bounded self', *Representations*, vol. 30, pp. 162–189.

Payne, JG 2016, 'Grammars of kinship: biological motherhood and assisted reproduction in the age of epigenetics', *Signs: Journal of Women in Culture and Society*, vol. 41, no. 3, pp. 483–506.

Petryna, A 2013, *Life exposed: biological citizens after Chernobyl*, Princeton University Press, Princeton.

Pitts-Taylor, V 2016. *The brain's body: neuroscience and corporeal politics*, Duke University Press, Durham.

Pottage, A 2002, 'Unitas personae: on legal and biological self-narration', *Law & Literature*, vol. 14, no. 2, pp. 275–308.

Pottage, A 2004, 'Introduction', in A Pottage & M Mundy (eds), *Law, anthropology and the constitution of the social: making persons and things*, Cambridge University Press, Cambridge.

Pottage, A 2007, 'The socio-legal implications of the new biotechnologies', *Annual Review of Law and Social Science*, vol. 3, pp. 321–344.

Rabinow, P 2005, 'Artificiality and enlightenment: from sociobiology to biosociality', in JX Inda (ed.), *Anthropologies of modernity: Foucault, governmentality, and life politics*, Blackwell, Oxford, pp. 179–193.

Rajan, KS 2006, *Biocapital: the constitution of post-genomic life*, Duke University Press, London and Durham.

Rees, T 2016. *Plastic reason. an anthropology of brain science in embryogenetic terms*, University of California Press, Berkeley.

Richardson, SS 2017, 'Plasticity and programming: feminism and the epigenetic imaginary', *Signs: Journal of Women in Culture and Society*, vol. 43, no. 1, pp. 29–52.

Richardson, SS & Stevens, H 2015 *Postgenomics: perspectives on biology after the genome*, Duke University Press, Durham.

Rose, N 2007, *The politics of life itself: biomedicine, power, and subjectivity in the twenty-first century*, Princeton University Press, Princeton.

Rose, N 2013, 'The human sciences in a biological age', *Theory, Culture & Society*, vol. 30, no. 1, pp. 3–34.

Rose, N & Abi-Rached, JM 2013 *Neuro: the new brain sciences and the management of the mind*, Princeton University Press, Princeton.

Schrader, A 2017, 'Microbial suicide: towards a less anthropocentric ontology of life and death', *Body and Society*, vol. 23, no. 3, pp. 48–74.

Selkälä, T & Rajavuori, M 2017, 'Traditions, myths, and utopias of personhood: an introduction', *German Law Journal*, vol. 18, no. 5, pp. 1017–1068.

Shildrick, M 2019, '(Micro)chimerism, immunity and temporality: rethinking the ecology of life and death', *Australian Feminist Studies*, vol. 34, no. 99, pp. 10–24.

Strathern, M 1988, *The gender of the gift: problems with women and problems with society in Melanesia*, University of California Press, Los Angeles.

Strathern, M 1999, *Property, substance and effect: anthropological essays on persons and things*, Athlone Press, London.

Strathern, M 2005, *Kinship, law and the unexpected: relatives are always a surprise*, Cambridge University Press, Cambridge.

Sunstein, CR & Nussbaum MC (eds) 2004, *Animal rights: current debates and new directions*, Oxford University Press, Oxford.

Teubner, G 2006, 'Rights of non-humans? Electronic agents and animals as new actors in politics and law', *Journal of Law and Society*, vol. 33, no. 4, pp. 497–521.

Van Beers, B 2017, 'The changing nature of law's natural person: the impact of emerging technologies on the legal concept of the person', *German Law Journal*, vol. 18, no. 3, pp. 559–593.

Van Wichelen, S 2016, 'Postgenomics and biolegitimacy: legitimation work in transnational surrogacy', *Australian Feminist Studies*, vol. 31, no. 88, pp. 172–186.

Vatter, M & De Leeuw, M 2019, 'Human rights, legal personhood and the impersonality of embodied life', *Law, Culture and the Humanities*, https://doi.org/10.1177/1743872119857068.

Warin, M, Kowal, E & Meloni, M 2018, 'Indigenous knowledge in a postgenomic landscape: the politics of epigenetic hope and reparation in Australia', *Science, Technology, & Human Values*, https://doi.org/10.1177/0162243919831077.

Wishart, D 2016, 'Puzzling out law's person', *Indiana Journal of Global Legal Studies*, vol. 23, no. 2, pp. 743–764.

Yngvesson, B 2002, 'Placing the "gift child" in transnational adoption', *Law and Society Review*, vol. 36, no. 2, pp. 227–256.

PART II

Evidencing Persons

CHAPTER 5

Translating Proof: Contested Illness, Radiation Exposure, and the Health Claims of Nuclear Test Veterans

Catherine Trundle

Introduction

How does biomedical knowledge intersect or diverge with legal knowledge? This chapter explores this question by examining the politics of proof within contested illness claim-making. Specifically, it explicates the quest for justice among New Zealand and British military veterans of United Kingdom (UK) nuclear bomb tests who seek compensation, recognition, and redress from their governments.[1] Such a case allows us to understand how emergent genetic knowledge transforms how we understand the body and illness through the lens of genetic personhood, and how these translate or fail to translate into legal understandings of injury and deservedness. It also reveals how genetics creates opportunities to transcend the ethical limits of epidemiological research and its post hoc focus, as well as its reliance of death certificates and deceased subjects.

C. Trundle (✉)
Victoria University of Wellington, Wellington, New Zealand
e-mail: catherine.trundle@vuw.ac.nz

M. de Leeuw and S. van Wichelen (eds.),
Personhood in the Age of Biolegality, Biolegalities,
https://doi.org/10.1007/978-3-030-27848-9_5

I argue that despite genetics' promise to reveal injury in new visceral, visual, and personalized ways, such knowledge does not easily translate within a legal system that continues to prioritize other ways of framing and understanding proof of the injured subject. To make this argument, I propose the notion of *translating proof*, which reveals the disjunctive knowledge systems at play that prevent genetic personhood in claimants from translating into legal culpability for the accused, at the same time that it explains why the state's assertions of innocence do not translate as valid knowledge for veterans as they seek to understand their own bodily suffering.

Background

In 1945 the US developed the first nuclear bomb, one powerful enough to end a global war. Fearing they would lose their status as a global power if they did not build their own nuclear weaponry, The British Government set about building an atomic bomb. As Winston Churchill phrased it, "We must do it, it's the price we pay to sit at the top table," (in Groot 2005, p. 221). Britain first set about developing its own atomic weapons in the outback and remote coastal regions of Australia between 1952 and 1956. Then, between 1957 and 1958, they developed the more powerful hydrogen bomb technology through atmospheric tests in the Pacific, above Christmas and Malden Islands in Kiribati, then a British protectorate. It took an immense amount of manpower to build those bombs, with just over 22,000 British armed forces personnel, 551 New Zealand Naval men, 14,000 Australian and 150 Fijian troops deployed to the testing sites to prepare the needed infrastructure, and execute and record the bomb tests. Indigenous populations in both Australia and Kiribati were "relocated and removed" to make way for these experiments (Arnold and Smith 2006; Crawford 1989, 2001).

The British government has always maintained that they sent personnel to the tests to help execute a complex techno-scientific experiment. Moreover, in a Cold War environment, they hoped to "indoctrinate" service personnel in the realities of nuclear war so they would be able to perform their duties under the shine of a mushroom cloud. Many test veterans argue, however, they were sent as human guinea pigs, as part of a covert government experiment to test the bomb's effects on human subjects and human health. The British Ministry of Defence has always contended that their nuclear program was conducted safely. Except for a few known exceptions, the servicemen were exposed to negligible, harmless amounts of radiation. They wield documents containing gamma radiation records and military safety protocols, epidemiological studies, and the expert options of radiobiologists to discount the men's claims and to prove that they have exercised a reasonable duty of care.

Many nuclear test veterans, by contrast, remember a much riskier, less ordered military operation that bears little resemblance to its official archival record. They have their own genetic studies, nuclear physicist supporters, legal experts, and archival documents that suggest an alternative telling of history. Test veterans mobilize narratives, documents, experiences, and evidence to tell a story of governmental neglect, indifference, and even a concerted cover-up.

Crafting Efficacious Proof

The questions at the heart of this chapter are, in situations where illness, suffering, and its causes are contested, what exactly constitutes *proof*? What are the diverse purposes that proof-making serves? What makes proof efficacious and influential in some contexts and what makes it fail in others? What are the political, epistemic, and narrative maneuvers required by claimants to build their own proof and make it count? Proof operates as a process and an assemblage rather than an object, I argue, at the heart of which are various acts of translation. In conflicts over illnesses, their etiologies, and their resolution, proof contains a range of ideas and actions. These include, among others, breach and disorder, embodied experience, claim, evidence, interpretation, authority, argument, moral reflection, and legal contestation. And each of these processes has a specific logic, goal, and set of associated practices. Translation is the dynamic process that links them, transforming one aspect of proof making into another, providing the movement within proof towards resolution or irreconcilability. For example, enacting proof in claims over contested environmental illness requires translating personal and collective experiences into evidentiary biomedical facts and personal narratives of harm, and translating evidence of harm into an ethical, relational assertion about the moral nature of human intentionality, responsibility, and blame. This process is not linear, tidy, or predetermined and can be unpredictable in how it unfolds.

Scaling up, there are multiple processes by which proof translates, mistranslates, or fails to translate across social spheres. To explore this process involves following proof capaciously to whichever social arenas it is required, and to think hard about the work involved in making proof travel or become stuck. Here, claimants must navigate questions of proof across scientific, legal, medical, bureaucratic, archival, and embodied spheres of knowledge and practice, and proof gets constituted differently in each domain.

Crafting proof that is socially legitimated and recognized, as well as politically persuasive, and that substantiates claims of culpability and deservedness thus requires claimants to work hard on multiple fronts. Within each social arena for which proof is required, they must learn how to stabilize proof through relating and translating the various processes that shore up proof into something solid and persuasive. And then they must learn to make any proof that they manage to stabilize translate in the right way, and to try to prevent it from translating in the wrong way, across diverse social domains. Reveals this way, it becomes clear that the tasks claimants of contested illnesses face in building a successful case for redress is a daunting, sometimes impossible one. But it is also a deeply creative one, in which they have multiple avenues for action.

Translation

Translation is never a neutral exercise. Talal Asad (1986) challenges us to consider the asymmetrical power dynamics of any processes of translation, and as Judith Butler (1997) shows, translating one's identity and experiences within a foreign or dominant discourse, to make oneself intelligible, often requires a type of subjective violence, of accepting particular roles and subordinate positions within that cultural space (see also Giordano 2008; West 2005).

Cris Shore (2005) also argues that translation involves an unequal politics of meaning-making. However, unlike Asad, he stresses that such a process is often dialogic, involving a conversation that pushes each party beyond their own habitual limits of understanding. For many scholars, translation can be a productive force that does not just transfer an idea from one location of meaning to another, but which fruitfully pushes the boundaries of knowledge. For Eduardo Viveiros De Castro, translation is a "controlled equivocation" between different perspectives that creates a productive misunderstanding, a recognition of distance that is a basic condition of any relation with an exteriority. Translation is "the difference between two things that grounds the relation," he argues "Translation becomes an operation of differentiation, a production of difference, that connects the two discourses to the precise extent to which they are not saying the same thing" (2004, p. 20).

In this generative mode, Walter Benjamin (1996 [1923]) proposes that translation is not simply a process of creating direct equivalence between things, but of a translation harmonizing with the "*intentio*" of an original idea/work, and creatively rendering it meaningful, lively and relevant in a new linguistic or social context. It is a supplement, not a reproduction. This is what Scott Montgomery, in his historical work on science in translation, calls the "offspring" that has the potential to transform both the translated and the translator (2000). Benjamin describes this as the "afterlife" of the thing translated. "For in its afterlife – which could not be called that if it were not a transformation, and a renewal of something living – the original undergoes a change" (1996 [1923], p. 256).

Benjamin is of course describing an effective translation carried out in good faith. Bad translations can do the opposite. They can reduce, render irrelevant, or erase the spirit of the work, person or idea being translated, and at worst they can purposefully belie or misrepresent the intent of the original (see Moon and Fenton 2002). But perhaps not all bad translations are bad, but a matter of perspective. After all, might the intention of a thing be up for debate? Or might the intention of a thing be suspect, and worthy of subversion and reinterpretation? Indeed, might there be times when it is not an act of betrayal but an act of courage to translate something in a new way?

Such an approach does not background the moral weave essential to much translation work (for example, Callon 1986; Latour 1987). As Susan Gal points out, "citation/translation can do... metapragmatic work: justification, accusation, alliance, mobilization. These effects, however, are invisible if the analysis of circulation stops with notions of flow, friction, travel, and movement" (2015, p. 231). Arguing that proof is an ethically complex process of translation moors it to the accusations of blame, culpability, and neglect that are so often essential to it. It requires us to take seriously the affective and political dimensions of translating proof and to understand the reason why people *do* proof in their own way and on their own terms.

Epidemiological Proof

The UK government's denial of responsibility has often rested upon a 1980s–1990s government-funded set of epidemiological studies by the National Radiological Protection Board (NRPB) (Muirhead et al. 2003). This concluded that UK test veterans' rates of cancers were no higher than a control group. This was a comprehensive study, with a huge sample, and many health researchers with whom I spoke commented on its rigorous design and execution. But the test veterans refused to accept it.

Why was the government sponsored study and its proof invalid to test veterans? The NRPB study was based on death certificates and mortality rates, which required test veterans' health to have failed in an absolute sense, in order for their bodies to count as evidence of illness and state culpability. This "inherently post hoc focus" (Shostak 2013, p. 191) reveals a medical rationale that is temporally retrospective, in that it engenders the past into the present. It creates a sacrificial relationality, by requiring the deceased portion of the group to become the means by which the "survivors" accounted for their own ill health. Many test veterans disliked that the sample of this study becomes more statistically in their favor the more of their friends died, and died suffering, and they resented what they saw as the implicit moral valance of a scientific method that made them useful to each other only when they were dead.

Many veterans also questioned why the study was so utterly detached from the *living* test veterans' ill bodies in the present and potentially the future. The study seemed unconcerned with present suffering, and its prevention. The surviving veterans were not called upon to offer biological or personalized evidence. The living veterans' bodies—those in need of medical care—remained unattended to by men they had never met, sitting in research centers, "just counting," as one participant Peter phrased it. Such studies offered no way through which their ongoing illnesses might become scientifically visible and treatable. They created only charts of numbers to represent the lives, emotions, and messy physical demise of deceased men. In other words, not only did they deny the living any comfort, but they did not ensure enough *witness* for the lives and trials of those test veterans who had died in their fifties and sixties. It was not the right type of proof of the right type of object with the right type of moral intent. As Diane Nelson (2015) argues, numbers are not inherently dehumanizing. But critiques of specific figures often argue that they have to count and account for the right thing, that they can produce unreason as much as reason, that they make certain realities visible at the expense of others, and that they often strategically mask the uncertainties of the what, how, and who of counting.

Peter pointed out that the military records upon which the study was largely based were notoriously inaccurate, no matter how rigorously the NRPB tried to crosscheck them. In the 1950s, rank and file soldier bodies just didn't count enough in military hierarchies and governmental budgets to require careful and individualized health surveillance. Test veterans challenged how accepting the study team was of government information when deciding who to include in their study sample. Some of the test veterans excluded from the sample for being deemed at no risk of radiation exposure could well have been exposed to radiation, Peter reasoned. It just wasn't indicated in military records that didn't accurately document where they were *really* posted and what their duties *really* involved. "They were awful record keepers, lost records, put the wrong stuff in, I've seen them. Why would anyone trust those documents?"

Many men I talked with who had requested their military records and military health records reported finding errors in them. One man was told definitely by the Ministry of Defence (MoD) that he was never at Christmas Island. He sent them a black and white photo of himself in front of a mushroom cloud. Veterans had learned, through first-hand experience, not to trust military archival information, and any evidence that was based upon this. And the researchers of the study, seeking scientific distance from the objects of their study, had never consulted individual test veterans to cross-check any information. The study invalidated test veterans' carefully built up community knowledge, and was based on information that veterans could not or would not trust.

"They thought they'd find nothing, so of course, they found nothing." Bill, a British test veteran said to me of the NRPB study. While I could find no evidence of this implied corrupt collusion, Bill's complaint opens a space for thinking through some of the central tenets of scientific methods—what they foreclose so as to produce ordered meaning. The NRPB study did rely upon a set of a priori facts. These acted as a lens to think through the possibilities at play. Such assumptions were, however, carefully produced. Epistemic weight existed behind them. They reflect the broader ways that scientific facts unfold relationally, as the legitimacy of new truths gets built out of the solidity of increasingly "black-boxed" facts.

The NRPB study argued that, based on the findings of previous studies of radiation exposure, particularly those in Hiroshima and Nagasaki, blood cancers are the only conditions that can be clearly linked to high dose radiation exposure, and normally emerge within 25 years of exposure (Muirhead et al. 2003). As a consequence, hematological cancer death would be a special focus in the study. Taking this as a given limited what the study could reveal with credibility to that which was already possible and imaginable in an established scientific field of vision. When the third follow up study found that liver, prostate, and bladder cancers were "significantly higher" in test veterans than in controls, the lack of similar findings in other epidemiological studies of radiation exposure and health resulted in the authors discounting these as a "chance findings." But test veterans argued that radiation research is beginning to show much wider effects beyond blood cancers, and radiation was not the only toxins to which they were exposed. Others included cigarette smoke, asbestos, various atmospherically dispersed chemicals from the exploded bomb components, pesticides, and even a bad diet. The still largely unknown synergistic effects of multiple exposures could potentially, test veterans argued, lead to a wider range of illnesses than science cannot currently measure (Trundle et al. 2014).

In the study, the authors used words such as "may," "likely," "reasonable," and "possibility" to describe their findings, and carefully laid out the limits to that which the study had analyzed, and the potential difficulties in both methodology and interpretation. The scientists themselves understood the ultimately reductive, tenuous nature of their findings. Despite this, the government treated the results as a more certain proof, declaring publically that it proved no harm had befallen the men due to radiation exposure. Politicians not only relied upon the study's specific findings to determine if the veterans' deaths were linked to their service, but extrapolated out further. They translated the lack of data about ongoing illnesses, morbidity, into proof that *no such* health problems existed. But this data had simply not been collected. In refusing to accept this epidemiological study, test veterans were thus rejecting its epistemic limitations and its political translation. Test veterans then turned to another field, that of genetic research, for the proof they desired.

Genetic Proof

Test veterans have focused on the emerging realm of genetic science because they believe it can enable a more flexible type of proof (Trundle 2017; Trundle and Scott 2013). It reveals biological vulnerability in the form of damaged genes, rather than manifest illness. It locates genes that are likely to increase the risk of illness more broadly in the present and future. Test veterans are thus attempting to transform the ways that bodies can provide legal evidence, and how suffering-as-proof comes to be legally redefined, from a manifest or completed process, a biological endpoint, to a situation of future potentiality and of latent risk. Such evidence places living test veterans' own bodies center stage, and as a consequence, their own fears and concerns about the trajectory of their health becomes more socially and medically legible. In the mid-2000s New Zealand test veterans commissioned a Massey University cytogenetic study (Wahab et al. 2008), which concluded that test veterans had three times the rate of chromosomal aberrations compared with a control group. Importantly to them, it showed *which* test veterans had high rates of aberrated cells.

Scale is fundamentally altered in the search for molecular biomarkers, as individuals, not just populations, come into view. Environmental epidemiology can make assertions about how many extra illnesses or excess deaths have occurred in a group due to a particular causal factor. These are above those deaths and illnesses naturally occurring in the wider nonexposed population (for example, an extra three in every thousand). What it cannot do is say whose specific deaths or illness are attributable to an exposure. Genetic research, by contrast, gets personal.

The cytogenetic study relied upon two very different assumptions from the NRPB study. The NRPB focused upon the possible effects of more acute levels of gamma radiation exposure. The genetic study, by contrast, accepted that low-level radiation exposure, in the form of ingested radioisotopes could continue to have long-lasting effects, for which evidence has begun to accumulate (for example, Hande et al. 2003). The second assumption of the Massey study was that people with high rates of chromosomal mutation suffer from worse health overall, for which evidence is also emerging. But the cytogenetic study did not measure either low-level exposure rates or the exact health effects of the test veterans' genomic instability.

When the test veterans presented their research to a government expert advisory panel, the panel accepted that the men had chromosomal damage, but their report to the Minister of Veterans Affairs declared that they could not conclude if this had resulted in ill health. The government made it clear that the type of proof they required would be bodies not simply at risk of illness, but manifestly, measurably ill or dying. In the end, genetic proof was not the right type of proof for the panel experts or the government.

One of the panel experts offered to do an epidemiological mortality and morbidity study for the veterans, but the head of the test veterans association refused, saying that would be another five years and they would all be dead by then. Once again, this type of proof was too temporally unresponsive, too cruelly detached, too prepared to let their health fail before anything needed to be done, too easily used by the government to wait a little longer before they acted. The veterans were trying to change the conversation about what, morally and biologically, proof needed to reveal, what action it needed to inspire, and how urgently proof-making needed to unfold.

Legal Proof

In 2008, just over 1000 test veterans (or their descendants) from the UK, Australia, New Zealand, and Fiji joined together in a Group Litigation Order seeking damages from the MoD (UK Supreme Court 2012). Injury, the claimants alleged, was due to a breach of duty and negligence on the part of the MoD in preventing radiation harm during the British Nuclear Testing Programme. Due to the 50-year delay between the nuclear tests and the litigation, claimants had to first prove to the British court that their claims were not statute barred. In other words, that they were legally entitled to make a claim so long after an injury. To test this, the defendants and claimants picked ten individual "lead cases" to test the questions of legal liability and limitation.

Thus in the preliminary court case to determine if the test veterans' case could go to full trial, an entirely different question of proof was at stake to that in the medical domain. Of significance was section 11 of the Limitation Act (1980), "Special time limit for actions in respect of personal injuries." It states that "any action for damages for negligence, nuisance or breach of duty…needs to be brought within three years." According to the Act, the time limit of three years begins from the date of the injury or harm, or the date of knowledge of the person injured. For nuclear test veterans, whose illnesses appeared often slowly and decades later, the moment when they gained knowledge of their ill health and its attribution to low-level radiation was significantly delayed, and has proven a legally complex issue to pinpoint. Of particular importance in the legal case, has been the question of the distinction between knowledge and belief (UK Government 1980).

An important legal principle here divides knowledge into two forms: "actual" and "constructive" Actual knowledge refers to concrete and tangible knowledge of "actual" or "material" facts. Constructive knowledge refers to the knowledge that one would be expected to pursue and acquire if one was prudent and exercised reasonable care. The MoD contended that the nuclear test veteran claimants had actual knowledge. When that might be difficult to prove, they argued that the test veterans at least had constructive knowledge—it was knowledge that they *should* have known about.

But test veteran claimants disagreed that they could or should have known they were affected earlier, and thus should have started legal proceedings years earlier. As one test veterans' widow, Rose, explained in her testimony in the high court in London, "Having been told so definitely by the Government that my husband did not have a case, I always thought there was nothing else I could usefully do…I never had any money to pay for lawyers, but the fundamental obstacle was the attitude of the Government which I felt I had no choice but to accept."

Veterans were suddenly in the position of trying to prove the exact opposite of what they had been trying to prove for years. While for decades they had been trying to tell doctors, and each other, and government officials that they were sure something was wrong, they now had to prove they had not been sure at all. The proof that had been so important in building up their own community, and in working with the media to pressure the government, mistranslated, or translated adversely, for them within the High Court and Supreme Court of Britain.

The government's lawyers pointed out that test veterans were active in the 1990s in talking to the media, and distributing newspaper clippings about their health concerns. The test veterans' legal counsel countered that, "The question arises of when belief becomes knowledge. Many of the claimants have applied for military disablement pensions … and at these the government said 'No you've not been exposed'. It's hopeless to say they had knowledge therefore." He asked, how could a belief amounted to knowledge, when the claimants were constantly being told that their beliefs were unreasonable. "They went to people who were competent and were told to go away."

However the majority of the presiding Lords of the Supreme Court disagreed, ruling that "If a claimant comes to believe that there is a causal connection between his condition and the matters complained of, it will matter not from where he derived that belief. If the belief were of such strength that it was reasonable to expect him to start investigating his claim, it would amount to knowledge" (UK Supreme Court 2012).

According to this logic, knowledge amounted to an internal state of mind, decontextualized and depoliticized, separated from a consideration of the odds of success, one's resources, legal impediments, or more pressing familial commitments. Knowledge was a stable, cumulative force that becomes "fixed" in a person at a precise moment based on an abstracted notion of conviction. It could not emerge and recede, and it was stripped of the pragmatics of hope, and realism or any feelings of despair and helplessness. By contrast for Rose and the other claimants, knowledge involved a broader context of unequal power relations that constrained both the possibilities for imagination, and the will to act.

Conclusions

This case-study reveals how proof and its varied logics, evidences, arguments and rationales do not convert across different social spheres with ease. The contests I have detailed here reveal arrangements and practices of knowledge, power, and authority within medicine, the law, state bureaucracy, and grassroots health social movements which preclude or enable particular modes of proof emerging, traveling, transforming, or being curtailed. As Viveiros De Castro argues, "Translation becomes an operation of differentiation, a production of difference that connects the two discourses to the precise extent to which they are not saying the same thing" (2004, p. 20). Sometimes this work of translation, this production of difference, worked in the test veterans' favor, such as when personal and collective stories about suffering were written down into newspaper articles or turned into testimony in pension appeal courts. Other times this translation worked against them, like when their newspaper stories made it back into the Supreme Court to haunt them, or when their genetic research did not provide the government with the right type of numbers or the right type of suffering.

Understanding processes of translation as intrinsic to the texture of proof and its efficacy helps us understand the processes that entangle citizens and state officials in intractable, polarized conflicts. This reveals the unequal arena into which claimants enter to make their case, with assumptions, rules, and processes often stacked against them being heard on their own terms.

The test veterans case reveals how the institutional power of state agencies is in part constituted through practices and structures that ensure the indecipherability and incompatibility of different aspects of the proof-making process to each other, which stalls the work of proof-making. They ask us to consider, when are personal accounts of the past acceptable as evidence for state agencies? When and how do state laws, such as the statute of limitations, render memories unreliable? And when does medical proof support or undermine legal proof?

This in/decipherability thus operates both within and across varied social domains. The state is not monolithic, made up of diverse agencies, actors and intentions. And it is this heterogeneous dimension that intentionally and unintentionally constitutes a core aspect of state power, as well as its susceptibility to critique and challenge. The plural logics of different state experts and agencies conflict with each other and often do not hold outside their sites of production, demanding distinct and incompatible questions and answers vis-à-vis the subject of contested illness, somatic harm, and state responsibility. This requires of claimants giddying amounts of knowledge across domains of expertise and an ability to transform how one speaks, thinks, argues, and acts as required. The complexity of such acts of translation, and the time required to become proficient in them, can overwhelm, wear down, or intimidate claimants.

Yet processes of translating proof also reveal the multiple spaces for action that claimants can carve out in a territory that soon becomes a familiar battleground not only for their claims, but for their sense of worth and purpose, for building rich claimant communities, and for crafting valued self-stories and life narratives. It is thus important not only to ask why claimants' proof fails to gain traction and recognition by state agencies, but also why the state's systems and objects of proof can be so resolutely rejected by claimant groups. This is partly because, as test veterans show, tenacious claimants can become skilled at playing the state at its own game.

The plural modes and questions of proof that test veterans encounter in various state institutions reinforce their sense that there is no singular, authoritative truth or evidentiary site upon which their claim rests. This does not mean that they or other claimant groups see all truths as equal, but rather, that they come to recognize that what is required to make a claim count involves the right type of proof, at the right time, in the right place, explained by the right person, in the right way. This realization only strengthens claimants' resolve to continually look for other, better types of proof and proof-making agents when the state's proof fails them. In short, if the state can do this, then so can they. In other words, the untranslatability of plural forms of proof within and across varied domains both bolsters the state and its experts' hold over truth claims, *and* reinforces opposition to it.

In such situations, claimants refuse to let state proof-making negate their own experiences and moral frameworks, and actively build alternative forms of proof and new ethical rationales for why proof is needed and what it must do. Or they translate "pure" science back to the state and medical officials as a morally freighted and politically unclean project. While they might not ultimately win their fight, test veterans are able to imagine proof as a different ethical project compared to many accepted models of medical and legal proof, and so, ultimately, they give us the opportunity to imagine proof in new ways. This may be what Benjamin (1996 [1923]) meant when he talked about translation as a form of renewal.

Note

1. Since 2009 I have conducted 70 interviews in New Zealand and the UK with test veterans, family members, solicitors, scientists, historians, advocates, and government officials. I have also collected documentary materials from the national archives of New Zealand and the UK, observed legal proceedings in the Appeal and Supreme Courts in Britain, and carried out participant observation within test veteran networks.

References

Arnold, L & Smith, M 2006, *Britain, Australia and the bomb: the nuclear tests and their aftermath*, Palgrave Macmillan, Basingstoke.

Asad, T 1986, 'The concepts of cultural translation in British social anthropology', in J Clifford & GE Marcus (eds), *Writing culture: the poetics and politics of ethnography*, University of California Press, Berkeley.

Benjamin W 1996 [1923], 'The task of the translator', in M Bullock & MW Jennings (eds), *Walter Benjamin, selected writings volume 1, 1913–1926*, Harvard University Press, Cambridge.

Butler, J 1997, *Excitable speech: a politics of the performative*, Routledge, New York.

Callon, M 1986, 'Some elements of a sociology of translation: the domestication of the scallops and the fishermen of St. Brieuc Bay', in J Law (ed.), *Power, action & belief: a new sociology of knowledge?* Routledge and Kegan Paul, London.

Crawford, J 1989, *The involvement of the Royal New Zealand Navy in the British nuclear testing programmes of 1957 and 1958*, Ministry of Defence Report, Wellington.

Crawford, J 2001, *New Zealand observers and indoctrinees at nuclear weapons tests: 1956–1958*, Ministry of Defence Report, Wellington.

DeGroot, G 2005, *The bomb: a history of hell on earth*, Pimlico Random House, London.

Gal, S 2015, 'Politics of translation', *Annual Review of Anthropology*, vol. 44, pp. 225–240.

Giordano, C 2008, 'Practices of translation and the making of migrant subjectivities in contemporary Italy', *American Ethnologist*, vol. 35, no. 4, pp. 558–606.

Hande, M, Azizova, T, Geard, C, Burak, L, Mitchell, C, Khokhryakov, V et al. 2003, 'Past exposure to densely ionizing radiation leaves a unique permanent signature in the genome', *American Journal of Human Genetics*, vol. 72, pp. 1162–1170.

Latour, B 1987, *Science in action: how to follow scientists and engineers through society*, Harvard University Press, Cambridge.

Limitation Act 1980, https://www.legislation.gov.uk/ukpga/1980/58.

Ministry of Defence (Respondent) v AB and others (Appellants) judgement 2012 U.K. Supreme Court, https://www.supremecourt.uk/cases/docs/uksc-2010-0247-judgment.pdf.

Montgomery, SL 2000, *Science in translation: movements of knowledge through cultures and time*, University of Chicago Press, Chicago.

Moon, P & Fenton, S 2002, 'Bound into a fateful union: Henry Williams' translation of the Treaty of Waitangi into Maori in February 1840', *Journal of the Polynesian Society*, vol. 111, no. 1, pp. 51–63.

Muirhead, C, Bingham, D, Haylock, R, O'Hagan, J, Goodill, A, Berridge, G et al. 2003, *Mortality and cancer incidences 1952–1998 in UK participants in the UK atmospheric nuclear weapons tests and experimental programmes*, National Radiological Protection Board, Oxon.

Nelson, D 2015, *Duke who counts? the mathematics of death and life after genocide*, Duke University Press, Durham.

Shore, C 2005, 'All in the translation: interpreting the EU constitution', *Sites*, vol. 2, no. 2, pp. 10–32.

Shostak, S 2013, *Exposed science: genes, environment, and the politics of population*, University of California Press, Berkeley.

Trundle, C 2017, 'Genetic bystanders: familial responsibility and the state's accountability to veterans of nuclear tests', in S Trnka & C Trundle (eds), *Competing responsibilities: the ethics and politics of contemporary life*, Duke University Press, Durham, NC.

Trundle, C & Scott, B 2013, 'Elusive genes: nuclear test veterans' experiences of genetic citizenship and biomedical refusal', *Medical Anthropology*, vol. 32, no. 6, pp. 501–517.

Trundle, C, Singh, I & Broer, C 2014, 'Fighting to be heard: contested diagnoses', in A Jutel & K Dew (eds), *Social issues in diagnosis: an introduction for students and clinicians*, Johns Hopkins University Press, Baltimore, MD.

Viveiros de Castro, E 2004, 'Perspectival anthropology and the method of controlled equivocation', *Tipiti: Journal of Social Anthropology of Lowland South America*, vol. 2, no. 1, pp. 1–22.

Wahab, M, Nickless, E, Najar-M'Kacher, R, Parmentier, C, Podd, J & Rowland, R 2008, 'Elevated chromosome translocation frequencies in New Zealand nuclear test veterans', *Cytogenetic and Genome Research*, vol. 121, no. 2, pp. 79–87.

West, P 2005, 'Translation, value, and space: theorizing an ethnographic and engaged environmental anthropology', *American Anthropologist*, vol. 107, no. 4, pp. 632–642.

CHAPTER 6

Paradigm Change, Law, and Persons: Producing Legal Responsibility for Pain

Seamus Barker

Introduction

Drawing on the philosopher Paul Ricoeur's theory of narrative, and of a narrative human subject, and inspiration from the philosopher of science Thomas Kuhn, this chapter will demonstrate how science and law can be coproduced specifically through the operation of "paradigms," in this case as they relate to pain. Rather than taking Ricoeur's writings on the law (2000) as a point of departure, instead I will take up his use, in the first two volumes of *Time and Narrative* (1984a, b), of the term "paradigm," using it to mean a framework from which narratives of particular happenings are derived. A paradigm—as a framework from which individual emplotments emerge—will be understood, as it is by both Ricoeur (1984a) and Kuhn (2012), as constituted not simply through a theory, model, or set of practices or texts, but also essentially constituted in the shared schema through which a community of human subjects perceive, act, and establish significance. More specifically, according to

S. Barker (✉)
Department of Sociology and Social Policy, University of Sydney, Sydney, Australia
e-mail: seamus.barker@sydney.edu.au

M. de Leeuw and S. van Wichelen (eds.), *Personhood in the Age of Biolegality*, Biolegalities,
https://doi.org/10.1007/978-3-030-27848-9_6

Ricoeur these processes are integrated through narrative *emplotment*—a term which here refers to an active, interpretative mode of understanding as much as any formal plot structure. In various writings on law (2000), narrative (1984a), and ethics (1992), Ricoeur also emphasizes the importance of the character, and the way that, through human action (in stories and in life), moral responsibility becomes enjoined to causality. In considering how legal responsibility for pain is produced, we will look to the way that nonhuman actants—such as an "injury"—are bound up in the same causal-moral economy as human characters, through processes of counterfactual inference. As well as these theoretical resources, this chapter will look at emplotments of selected legal narratives—taking the form of judgments in case law relating to workers' compensation disputes in the Australian State of Victoria—as well as aspects of the paradigm from which they will be understood to emerge, particularly as it is constituted by Victoria's *Workplace Injury Rehabilitation Compensation Act 2013* (*WIRCA* 2013). More specifically, we will look at the requirement under the *WIRCA 2013* to "disentangle" psychological or behavioral pain and suffering consequences from physical pain and suffering consequences, as these arise from a single physical "serious injury."

According to Ricoeur (1984a, p. 69), narratives always involve characters and, like tragedies, have a typical "form" in which a situation is unexpectedly transformed for the worse, producing suffering. For the plot to remain coherent, such "reversals" must be accounted for, through the establishing of causal and teleological interconnections between narrative elements, which creates a sense of narrative necessity: the sense that, once certain events had transpired, things could not have turned out any other way (Ricoeur 1984a, pp. 43–44). For Ricoeur (1984a), narrative emplotment is not a strictly authorial process. Rather, emplotment is completed through the interpretative work of an audience, who must actively grasp together interconnections that are only implied in a text. Ricoeur (1992) further uses this model of emplotment to ground a narrative version of the human subject that is understood to emplot, rather than freely author, the past events and sufferings of their own life into a (usually) coherent narrative. Given this fundamental form, narratives, according to Ricoeur (1984a, p. 56), do not present mere material cause and effect, rather they emplot human "acting and suffering." "Acting," in this context, implies an actor that has ownership of—responsibility for—an act, but such "ascription" of an action to an agent is not simply internal to narrative syntax. Rather, action and suffering can be made intelligible to a person through their "preunderstanding" (Ricoeur 1984a, p. 64). This "preunderstanding" is disclosed, at a general level, through a preexisting cultural, symbolic, and also normative framework. In this way, the fact that narrative "acting and suffering" are intelligible to a community is inseparable from the way that "acting and suffering" demand moral evaluation, according to Ricoeur. The law, of course, provides an explicitly normative framework and, as we shall shortly see, the *WIRCA 2013* provides a distinctive framework for emplotting physical injuries in terms of action, suffering, and the "imputation" of causal-moral responsibility. Scientific and medical paradigms of pain also provide general frameworks—models of pain that aim at universal applicability—from which emplotments of individual instances of pain, in the form of causal explanations, can be derived. We can consider whether scientific and medical emplotments necessarily become figured in terms of "action and suffering," and thereby inescapably demand moral evaluation, or whether they are more neutrally translated into the framework of the law, and only there acquire a normative dimension. If sequences of "action" demand moral predicates, as Ricoeur insists, we can consider, too, whether a paradigm—existing as potential, rather than realized, emplotment—might have what Habermas calls a "crypto normative" function, arising from a concealed moral structure. Accordingly, as we turn now to consider workers' compensation in Victoria in more detail, we can consider whether, for all of the commitment of science to objectivity, a crypto normative structure embedded in a paradigm of pain might function as a powerful moment in the coproduction of science and law, and help explain the pertinacity of body–mind dualism.

"Serious Injury" Claims and Pain and Suffering

There are various statutory entitlements under the *WIRCA 2013*, similar to those of jurisdictions in other welfare states, that relate to any injury occurring at work in Victoria, including the coverage of medical expenses and a capped percentage of lost wages for up to two years, if the injury requires treatment or materially contributes to an incapacity for work (Churilov 2017). Legal disputes can arise when an insurer closes a claim, which typically occurs on the basis that they find the worker to have no incapacity for work, or, where incapacity for work exists, they find that it is not materially contributed to by the claimed injury. Here, we will focus on a different type of case—applications for "serious injury"—since our findings will remain pertinent to the broader picture of workers' compensation law in Victoria, and to our broader interest in the coproduction of science and law via the operation of paradigms.

A central feature of the paradigm from which workers' compensation legal judgments are emplotted is the *WIRCA 2013*. This *Act* stipulates that someone injured at work can bring a Common Law damages suit for pecuniary (economic) and/or nonpecuniary losses, although first they must satisfy the court that it was a "serious injury" and then prove negligence on the part of the wrongdoer (2013, s. 318–327). In practice, a significant proportion of workers' compensation cases heard in Victoria relate to making such a determination regarding "serious injury." The *WIRCA 2013* defines "serious injury" as "(a) permanent serious impairment or loss of a body function; or (b) permanent serious disfigurement; or (c) permanent severe mental or permanent severe behavioral disturbance or disorder" (2013, s. 325). A dichotomizing logic, in which injuries are understood as either physical or mental/behavioral, thus is immediately apparent. This logic is made more clear when we note that in a serious injury application for bodily impairment, "the psychological or psychiatric consequences of a physical injury" are not to be taken into account, while in a serious injury application for a mental or behavioral impairment, similarly "the physical consequences of a mental or behavioral disturbance or disorder" are not to be taken into account (*WIRCA 2013*, s. 326). As a result, to satisfy the court that a serious physical injury has occurred, the *WIRCA 2013* specifies that an injured worker should undergo formal assessment of impairment, according to the *American Medical Association's Guides to the Evaluation of Permanent Impairment 4th Edition*, and reach a 30 percent Whole Person Impairment (WPI) rating, but that any scores arising from assessments in the chapters in the *Guides* on chronic pain (Chapter 15), and on psychological and psychiatric impairment (Chapter 14), must not be included in the tally. These exclusions exist even though the *WIRCA 2013* specifies that "serious," in the context of a workplace physical injury allegedly resulting in nonpecuniary losses, means that the court must "be satisfied by reference to the *consequences to the worker* of any impairment or loss of a body function… with respect to… *pain and suffering*" (s. 325; emphases added). In the *WIRCA 2013*, then, we can already begin to discern an implicit framework for the preunderstanding of "action," including a moral dimension, as it relates to pain and injury. To see more clearly the relationship between this legal paradigm and the emplotments that emerge from it, let us consider an alternative pathway that developed for serious injury applications: what has become known as "the narrative test." Not only is this salient to our interests but, as Tsalamandris and Lombard (2011, p. 37) note, since very few injured workers are evaluated has having a 30 percent WPI rating, based on physical impairment alone, the "narrative test" has become the most common pathway for successful application for a certificate of a "serious injury," as a requirement to proceeding to a Common Law claim for damages arising from a workplace injury.

In *Humphries v Poljak* (1992), a Victorian Supreme Court case, Crockett and Southwell JJ determined that if the plaintiff could satisfy the court through their testimony that they indeed had suffered a serious injury, then they could proceed to recovering their losses, setting a precedent for the so-called "narrative test." In *Meadows v Lichmore Pty Ltd* (2013), in the Victorian Supreme Court of Appeals, Maxwell ACJ, Robson and Dixon AJJA interpreted the significance of the stipulation, in the *WIRCA 2013*, that psychological consequences of a physical injury must be disregarded for the purposes of determining if a physical injury is "serious," in the context of this "narrative test":

> In the present case, there was a physical injury. Mrs Meadows had also developed a "pain syndrome" or functional overlay. The resultant pain and disability therefore had both physical and psychological aspects… the applicant will need to take the next step and "disentangle." That is, the applicant will need to be able to separate the physical contribution to the pain and suffering from the psychological, in order to be able to satisfy the Court that the pain and suffering consequences attributable to the physical injury satisfy the statutory test.

The requirement to "disentangle," through the process articulated above, now is standardly applied if a court has reason to believe that there might be psychological or behavioral elements contributing to pain and suffering consequences of a physical injury. This scenario would be argued for by the insurer's legal team, typically relying upon evidence from various specialist medical assessments organized and paid for by the insurer, which the worker, under the *WIRCA 2013*, must submit to. At this point, we can try to understand the logic behind "disentanglement," and whether such logic reveals a moment of the coproduction of science and law. In seeking a possible basis for disentanglement, let us turn now to different scientific and medical paradigms of pain.

Paradigms of Pain

Cohen et al. (2013) note that discoveries in sensory physiology at the end of the nineteenth century produced knowledge of nociception—the process by which noxious stimuli are encoded and transmitted to the brain—which led to the idea that pain was a raw sensation, proportional to, and predicted by, a noxious stimulus, such as commonly provided by an injury. This scientific model was deployed in medicine such that, there, pain was understood as a symptom of a disease. A theory of pain as a fixed transmission system could not account, though, for the widely observed modification of pain by context and emotion, and so any pain that could not be fully accounted for through evidence of a proportionately severe injury, or disease, was classified as of "non-organic origin" (Cohen et al. 2013, p. 1285). In the middle of the twentieth century, complaints of pain of putatively nonorganic origin could be classified psychiatrically as "malingering" (deliberate deception regarding symptoms), "hypochondriasis" (exaggerated interpretation of symptoms leading to fear of a serious disease), or "psychogenic pain disorder" (unresolved psychological conflict or unconscious motive converted into physical pain) (Barker 2017, p. 395). These psychiatric constructs, along with a mechanistic medical model, formed a dualistic paradigm of pain (Barker and Moseley 2017), which we can relate to the requirement in the *WIRCA 2013* to keep "organic" and "non-organic" pain and suffering distinct.

Cohen et al. (2013, p. 1285) suggest that the idea of pain of "non-organic origin" gradually became unacceptable, and was superseded in 1965 by the Gate Control Theory of pain, which posited that that the signals that encode noxious stimuli—nociceptive signals—can be modulated at the spinal cord, such that the nociceptive signal that ultimately arrives at the brain can be greater or lesser than that originating in the periphery. An implication of this theory is that tissue damage does not predict pain, and that it is normal, not psychopathological, to sometimes experience pain that is disproportional to an injury. Indeed, the very idea of proportionality between injury and pain becomes dubious.

Taking these ideas even further is a theory developed by one of the authors of Gate Control. According to Melzack (2001, pp. 1379–1380), a signature pattern of neural activity—a "neuromatrix"—gives rise to the conscious experience of a stable "body-self." This means that the brain maintains a continuous representation of the body, which ideally maps closely on to the physical body. Additionally, the body-self neuromatrix helps the organism maintain homeostasis by activating responses to threat, including hormonal and immune responses but also movement, communication or other strategic action (Melzack 2001). Such "outputs" from the body-self neuromatrix are responsive to a range of "inputs" (Melzack 2001, p. 1382). Crucially, in the neuromatrix model, pain is not one of these *inputs to* the brain, but rather is one of the *outputs of* the brain. Specifically, nociception (not pain) can be an input, along with thoughts, memories, and feelings, as well as visual, auditory, kinesthetic, and somatosensory inputs (Melzack 2001) If these inputs suggest that homeostasis is threatened, pain can consequently be produced as a spur to action. Within this paradigm, the "normal" pain one can experience in a limb is not so different, in certain respects, to the phantom limb pain that an amputee can experience. In neither case is the pain "in" the actual body and in both cases the pain is "in" the body-self: the virtual body that is based on the brain's processing of available information (although mismatched with the physical body in the case of a phantom limb). Accordingly, we can understand the neuromatrix theory of pain to be non-dualistic and thereby *incommensurable* with a dualistic paradigm of pain, although we will later unpack this incommensurability, inspired by both Kuhn (2012) and Ricoeur (1984a) in terms of the potential for emplotment of causal explanations of pain.

Cohen et al. (2013, p. 1286) also note that a line of research that grew out of Gate Control Theory discovered a process that can occur with chronic pain that came to be known as "central sensitization." "Central sensitization" involves various neuroplastic changes in the structure and function of the central nervous system, which can amplify nociception at the spinal cord and make pain occur much more readily and intensely in a regularized fashion, regardless of context. These processes can be demonstrated in laboratories, but not through diagnostic imaging tests for people with chronic pain. We will later consider whether "central sensitization" aligns with a dualistic or a non-dualistic paradigm of pain.

In addition to these models of pain, there is also a vast body of research into pain-related disability. Certain cognitive and behavioral constructs, such as "kinesiophobia" (fear of movement when in pain), "pain catastrophizing," self-efficacy when in pain, as well as mood disorders, have been found to be not simply associated with increased disability for chronic pain populations, but to exist in complex, multidirectional relationships with pain itself, thus breaking down any apparently neat division between organic pain, and nonorganic pain behavior (see, for example, Meints et al. 2019).

Understanding "Disentanglement"

Given the evident obsolescence in research settings of a dualistic paradigm in which pain is considered either "organic" or "non-organic," we can turn now to jurisprudence to try to understand why the *WIRCA 2013*, and the associated "narrative test," insist on "disentanglement" of psychological from physical pain and suffering consequences of a physical injury. To engage with this question we can consider aspects of legal reasoning, particularly relating to causation and legal responsibility. As Mendelson notes, "in the modern tort of negligence," which includes damages claims arising from "serious injury" in the workplace in Victoria, "the legal issue of causal connection between the defendant's conduct and the harmful result is … split into two questions … 'factual' causation … and 'proximate causation'" (2002, p. 66) Factual causation is determined by the "but for" test—which asks whether some harmful outcome *b* would have occurred if not for prior action (or omission) *a*—such that a nonexistent (counterfactual) world must be imagined, in which prior action *a* did not take place. Determining "factual causation," Honoré suggests, is "an essential tool for holding people responsible for the change they make in the world," but, however, this model must be delimited, since in a causal chain there can be innumerable prior causes and subsequent effects that are connected by necessity, but most are too remote from the ones at issue to reasonably be considered in terms of legal responsibility (2002, p. 9). Thus, "factual causation" (also called cause-in-fact) is necessary, but not sufficient, for establishing "proximate causation," (also called "legal causation," or cause-in-law). Establishing "proximate causation," in tort law, involves making a normative legal judgment that determines where legal responsibility for effects subsequent to a negligent act should end. *Novus actus interveniens* is a legal principle that refers to a new act understood to break the chain of continuity between a negligent act and a later harm (Honoré 2002). Traditionally, it has been understood that such an intervening act could be enacted by a third party, or by the claimant themselves, or as an "act of God." Honoré gives the example that someone who negligently causes a car crash that injures another person, who then requires hospital care, would not be responsible if the hospital was to be struck by lightning that in turn created a fire, burning the crash victim, even though the burns would not have occurred "but for" the driver's negligence (2002, p. 11). Similarly, if we modify that example, if the person injured in the initial crash then drove themselves to hospital, but was speeding there and created another crash causing further injury, this act also would break the chain with the original act and limit the original driver's liability. Finally, the intervention of a third party, for example if the injured driver was to be assaulted while in the hospital causing further injury, also can break the chain of responsibility for the original injurious party. Not coincidentally, under the *WIRCA 2013* workers are not covered for injuries that are deliberately or willfully self-inflicted or that arise from serious misconduct, which also would seemingly break the chain of responsibility extending from the employer (Churilov 2017). This mode of causal reasoning follows a strongly narrative logic, relying on emplotment of action and "imputations" of causal-moral responsibility to individual actors.

Bearing these principles in mind, it is worth considering that, if in the *WIRCA 2013* there was no statutory requirement to disentangle psychological from physical pain and suffering consequences, then it seems, in the aforementioned case of Mrs. Meadows, who after a "physical injury" subsequently "developed a pain syndrome or functional overlay," that the "but for" test of physical causation would be satisfied, in terms of the psychological consequences not occurring *but for* the prior physical injury (*Meadows v Lichmore* 2013). We should note, at this point, that the "egg shell thin skull" rule also applies in tort law, which means that if the victim of an act is unusually vulnerable (such as by having a skull as thin as an egg shell), the tortfeasor still remains responsible for damage more extensive than might usually arise from the harmful action (Klimchuk 1998, p. 124). The tortfeasor, in other words, must take their victim "as they find them." The "egg shell thin skull rule," we should note, is supposed to apply to psychological as well as physical vulnerability (Klimchuk 1998, p. 126). We should note, too, that since the case of *The Wagon Mound*, limitations to liability are often cashed out in terms of reasonable foreseeability (a doctrine which has a similar function to *Novus actus interveniens*, since new intervening acts that produce further harms will not have been foreseeable to the originally negligent party) (Rowe 1977). This criterion of foreseeability does not cancel, though, the "egg shell thin skull rule," since disproportionate harm that arises directly from a negligent act, even if contributed to by special vulnerability that was not foreseeable, still is the responsibility of the tortfeasor (Klimchuk 1998, p. 125). If no statutory requirement to disentangle existed, then, and Mrs. Meadows was "taken as she is found," her vulnerability to subsequently developing a "pain syndrome or psychological overlay," even if atypical, should not limit the liability of a party that had negligently caused her initial injury. Consistent with the "egg shell thin skull" rule and *Novus actus interveniens*, it seems that (if no statutory requirement to disentangle existed) only a new intervening act—for which responsibility could be imputed—could break the chain and reduce the tortfeasor's liability, in terms of the full extent of psychological, as well as physical, pain, and suffering arising from the original injury. *Novus actus interveniens* is not formally invoked in the *WIRCA 2013*, but I would suggest that it could provide an implicit, normative justification for the statutory requirement to disentangle. Could the development of "non-organic" pain and suffering consequences be figured so as to constitute a new intervening act, that breaks the chain of responsibility?

"Moral Hazard" as Imputed Agency

It is important to remember that a "serious injury" is defined in the *WIRCA 2013* in terms of its consequences for the worker, including pain and suffering. If we begin with this final step in the chain of action, and work backwards, and accept that psychological pain and suffering, such as described in *Meadows v Lichmore*, would not have occurred *but for* an initial injury, then we must ask how "disentanglement" is normatively justified. Even if a dualistic model of pain indeed was descriptively (scientifically) accurate, normatively there should still be no need to disentangle psychological from physical pain and suffering, using the "egg shell thin skull rule" as a normative guide, unless psychological pain and suffering, in a given instance, would have occurred even without any initial physical injury (which seems extremely unlikely). I would therefore suggest, in line with *Novus actus interveniens*, that the statutory requirement in the *WIRCA 2013* to "disentangle" has a normative basis only if "psychological pain and suffering consequences" are taken to constitute an intervening act, which breaks the chain of responsibility. Such intervening acts, according to Klimchuk (1998, p. 115), following Hart and Honoré, need to be "free" rather than "unfree," thus binding responsibility only to intentional actors, which conforms with Ricoeur's model of ascription. A complaint of pain and suffering arising from *malingering*—deliberate deception regarding symptoms—would clearly constitute an intervening act that breaks the chain of responsibility extending back to a negligent tortfeasor, but I would suggest that, under the *WIRCA 2013*, "non-organic" symptoms that are not "free" or voluntary are similarly framed as the responsibility of the sufferer.

To understand why this might be the case, it is important to consider Stapleton's argument that establishing "proximate cause," including where it is limited by *Novus actus interveniens*, is ultimately *not a causal analysis* that establishes the genuinely proximate cause of a harm, so much as a *normative judgment of fairness* regarding how and to whom legal responsibility should be imputed (2002, p. 18). Pain and workers' compensation involve vast amounts of money, of course, with workplace injuries costing Australia around four percent of its GDP (Safe Work Australia 2016). The *WIRCA 2013* acknowledges that it is normatively guided by the objective of not only providing fair compensation to injured workers, but of keeping the scheme in Victoria financially viable, including by reducing costs to insurers and employers, and detecting and deterring fraudulent claims (Churilov 2017). Yet, when it comes to "serious injury" applications, it is through the causal language of disentangling physical and psychological pain and suffering that this principle of limiting liability seems to be enacted. We can consider here an insurance concept called "moral hazard," which assumes that insurance against some event will increase the incidence of that event among the population so insured, or, put more generally, that incentives or disincentives built into a system will distort outcomes at a population level (Barker 2017, p. 396). Both deliberate malingering and "non-organic" "pain syndromes" are theoretically susceptible to "moral hazard" (Barker 2017, p. 396) in a way that verifiable physical injuries are not, and I would suggest that this susceptibility ultimately is the reason that "non-organic" pain and suffering, as much as intentionally fraudulent behavior, is figured as a new intervening act. In other words, the assumed distortion of outcome, at a population level, by financial incentive, as described by the actuarial logic of "moral hazard," becomes transfigured, in law, as individual agency—as it is the worker who apparently "acts" in developing the psychological "pain syndrome" that contributes psychological pain and suffering consequences.

Thus, I would suggest that, regarding those injured workers who present to insurers increased risk, it is through the figuration of the worker as the agent causally—and thereby morally—responsible for their own state of pain and suffering, and thus not covered for their losses, that the principle of fiscal restraint, as one of the normative bases of the *WIRCA 2013*, is enacted.

The Coproduction of Dualism

I have argued that it is a principle of fiscal restraint (which effectively shifts costs from employers and insurers to workers), and the transmutation of the risk of "moral hazard" into apparently individual agency, that stand behind "disentanglement." Yet, it is a dualistic paradigm that makes possible such imputations of responsibility for "non-organic" pain and suffering, which are narratively figured in law at the level of "action." Is this dualistic paradigm a function of medicine or science, or is it internal to the logic of a larger legal paradigm? If we turn again to the law in Australia, a Victorian Supreme Court appeal, *Heard v Bronzewing Linehaul Pty Ltd* (2013), was lodged on the basis that the plaintiff's pain and suffering consequences were indeed, and in contradistinction to the trial Judge's findings, organically, rather than nonorganically based. In the original case, the plaintiff's legal team emphasized the expert testimony of a pain specialist who diagnosed the plaintiff with "central sensitisation," which he described as a "scientifically proven organic change in the central nervous system pain pathways in reaction to the original injury." The Judge, though, saw the central sensitization explanation as an outlier among a range of medical opinions, including those provided by doctors engaged by the defense, such as orthopedic surgeons who found, in the words of the Judge, "no physical or radiological explanation for the appellant's current complaint of pain and no objective evidence of [nerve compression]" such that "from a physical point of view there was no reason for the appellant to be off work," as well as Occupational Physicians who found the appellant "fully recovered" from the physical injury and fit for fulltime work duties, though in need of "psychological counseling and cognitive behavioural therapy" for depression, and also by psychiatrists who diagnosed "adjustment disorder." The bench agreed with this earlier finding, and dismissed the appeal. In *McAleer v Austin Health* (2015), a Victorian County Court case, the very same pain specialist was drawn upon, and he again provided a report detailing central sensitization and its specifically organic underpinnings. In this case, the Judge took this pain specialist's testimony as the most persuasive and accepted the organic nature of the plaintiff's pain complaint. These cases reveal that when a modern model of pain such as "central sensitization" appears in the workers' compensation paradigm in Victoria, it must be rendered into a dualistic framework and figured explicitly as an "organic change." Even then, "central sensitization" is treated with some circumspection, and can be dispreferred to medical accounts that equate organic pain strictly with tissue damage verifiable by radiological imaging, and blame severe symptoms, when adjudged disproportional to such tissue damage, on psychological or psychiatric problems.

In law, the "but for" test of causation imagines the effects in a world where a particular action (that took place in the real world) has not taken place. We can begin to see that, in the case of a dualistic framework for pain operating within the Victorian workers' compensation paradigm, this process of counterfactual reasoning is more rigidly overdetermined. If we recall the bench's judgment in *Meadows v Lichmore*, as they sought to determine responsibility for Mrs. Meadows' pain and suffering, there a "pain syndrome" or "functional overlay" is figured as an actant that provides a "psychological contribution" to Mrs. Meadows' "pain and suffering consequences." At the same time, the "injury" is also figured as an actant that provides a "physical contribution" to Mrs. Meadows' "pain and suffering consequences." Relatedly, in *Heard v Bronzewing Linehaul Pty Ltd*, "central sensitization" must be framed, when introduced by the plaintiff's legal team, as occurring "in *reaction to* the original injury," such that "the injury" can remain the narrative actant and "central sensitization" thus cannot be figured so as to break the chain of the tortfeasor's responsibility. In this legal paradigm, then, to imagine pain and suffering without a physical cause is necessarily to imagine a mental or behavioral cause for that pain and suffering. To imagine pain without a mental cause is necessarily to imagine a physical cause for that pain. I would suggest that this overdetermined process of counterfactual inference is inseparable from what Ricoeur (1984a, p. 64) calls narrative "preunderstanding"—a culturally, symbolically, and normatively mediated framework—that renders "action" intelligible, including through an explicitly moral valuation, and makes emplotment possible. In this way, for an "injury" to be established as the actant that brings about pain is, by definition, for the sufferer to be absolved of causal-moral responsibility for that pain. To have pain that is not fully accounted for by an "injury" is, by definition, for the sufferer to be imputed causal-moral responsibility for that pain, via the mediation of a "pain syndrome" figured as both resulting from the sufferer's "action," and in turn acting to produce "non-organic" pain and suffering.

It is not coincidental that such a strict dualism of pain reproduces the dichotomizing logic of the law and the judgments it hands down. Terms relevant to the alternative, non-dualistic paradigm of pain—"Gate Control Theory" and "neuromatrix"—do not appear in searches of Victorian legal databases, and so presumably have not appeared in evidence. If such non-dualistic models were introduced in medical testimonies, in "serious injury" applications, to explain pain and suffering, they would prove incommensurable, I would suggest, with the narrative preunderstanding shared by those working within the paradigm of the *WIRCA 2013*, through which action is made intelligible according to dualistic terms. In the non-dualistic Gate Control Theory, a circular feedback loop between brain and spinal cord exists, in which each is reciprocally and continuously both acted upon and acting, thus challenging the linear structure required by legal narratives that have a beginning, middle, and end, and simple actant structures that maintain an unambiguous sense of legal responsibility for an instance of pain. Similarly, if the neuromatrix model understands pain as an emergent phenomenon arising from a complex system, with an enormous raft of networked structures and processes ultimately contributing to the "inputs" available to the brain, then it becomes difficult to emplot all of these as actants and actions, particularly where causes and effects converge, and indeed the notion of emergence is seemingly incommensurable with an actantial model. The neuromatrix model understands pain as an *output of* the brain, rather than as an *input to* the brain, and I would suggest this point of distinction indicates a true incommensurability between a non-dualistic paradigm of pain and dualistic legal paradigm. We should recall that, for Ricoeur, emplotment is finalized by the reader, who draws on their preunderstanding of action to clarify causal interconnections between narrative elements, as well as their moral significance. If the neuromatrix was used in medical testimony to explain a worker's pain, this emplotment could be (mis)interpreted through the law's dominant framework, with "the brain" figured, according to a dualistic preunderstanding, as the actant that produces "non-organic" pain (despite this being nonsensical from the position of the non-dualistic paradigm), thereby establishing the sufferer as the agent of their own pain. Indeed, such a (mis)understanding of action and responsibility seems particularly likely to happen given that pain—as defined by these non-dualistic models—can be distortable by incentives and disincentives (just like malingering or "non-organic" pain), and so always poses a fiscal risk, according to the actuarial logic of "moral hazard."

Conclusion

Paradigms can be understood as frameworks through which narrative emplotments, including those that feature causal explanations of particular instances, emerge. These emplotments are intelligible through a pre-understanding of action and suffering, involving human and nonhuman actants. Yet, rather than a narrative being merely a secondary derivation from a paradigm, we have seen that it is in emplotment that not only an ordering of cause and effect, but a particular moral economy, become realized. Ricoeur suggests there is a difference between "ascription" and "imputation," since the ascription of an action to an agent lacks the moral evaluation bound up in the term "imputation" (2000, p. 21). Yet, as Hart suggests, "action" itself is not transparently descriptive, but is always a "social concept and logically dependent upon accepted rules of conduct" (1949, p. 189). Siding here with Hart, our study of the disentanglement, after a physical injury, of psychological and behavioral pain and suffering from physical pain and suffering similarly reveals that the nouns and verbs by which actantial structures are classified, bounded, and ascribed action are not without moral force, even—especially—when they are supposed to be detached and clinical. This "crypto normative" operation occurs partly due to the counterfactual processes of reasoning that occur in narrative understanding, which mean that even nonhuman actants like an "injury" can be bound up in a causal-moral economy, through their potential to absolve human actants of moral responsibility, or when an actant like a behavioral "pain syndrome" is narratively configured so to break the chain of responsibility that began with another injurious party.

Different potentials for emplotment into action and suffering—including as these different emplotments produce different moral structures—mark different paradigms as incommensurable. Narrative provides a reduced framework, suited to action sequences that concentrate cause and effect on to presumed containers of causality, while being seriously restricted when required to account for emergent, nonlinear, networked processes and system dynamics, as we find in a non-dualistic scientific paradigm of pain.

As well as incommensurability between paradigms, we can see a single paradigm—body–mind dualism—functioning through the coproduction of law and medicine. Dualistic medical authority is not unilaterally translated into a passively waiting legal system, though. Instead, it is the dichotomizing juridical function of the law, and its normative judgments regarding allocations of responsibility for harms—which are based ultimately on notions of fairness, or the expression of powerful interests and actuarial logics, but not on physical causation—that authorizes dualistic emplotments in the law, which in turn legitimates broader dualistic paradigms, including as they persist in medicine. Thus, this chapter argues that paradigms of pain, existing as potentials for emplotment, occult underlying moral structures. The moral structure of a dualist paradigm—realized in emplotment and potentiating the coproduction of medicine and law—perhaps is one of the more forceful features if we try to explain why a shift in society's understanding of pain, from a dualistic to a non-dualistic paradigm, has not yet occurred.

References

Barker, S 2017, 'Subject to pain: Ricoeur, Foucault, and emplotting discourses in an illness narrative', *Subjectivity*, vol. 10, no. 4, pp. 393–410.

Barker, S & Moseley, GL 2017, 'Narrative coherence and medical explanations of psychosomatic pain', in CA Farkas (ed.), *Reading the psychosomatic in medical and popular culture*, Routledge, London.

Churilov D 2017, *Personal injury compensation in Victoria*, 2nd edn, Wolters Kluwer, Sydney.

Cohen, M, Quintner, J & Buchanan, D 2013, 'Is chronic pain a disease?', *Pain Medicine*, vol. 14, pp. 1284–1288.

Hart, HLA 1949, 'The ascription of responsibility and rights', *Proceedings of the Aristotelian Society*, vol. 49, pp. 171–194.

Heard v Bronzewing Linehaul Pty Ltd (2013) Victorian Supreme Court of Appeals 20.

Honoré, A 2002, 'Principles and values underlying the concept of causation in law', in I Freckelton & D Mendelson (eds), *Causation in law and medicine*, Ashgate, Burlington, pp. 3–14.

Humphries and Another v Poljak (1992) 2 Victorian reports 129.

Klimchuk, D 1998, 'Causation, thin skulls and equality', *Canadian Journal of Law and Jurisprudence*, vol. 11, no. 1, pp. 115–141.

Kuhn, T 2012, *The structure of scientific revolutions: 50th anniversary edition*, University of Chicago Press, Chicago.

McAleer v Austin Health (2015) VCC 2.

Meadows v Lichmore Pty Ltd (2013) VSCA 201.

Meints, S, Mawla, I, Napadow, V, Kong, J, Gerber, J, Chan, ST et al. 2019, 'The relationship between catastrophizing and altered pain sensitivity in patients with chronic low-back pain', *PAIN*, vol. 160, no. 4, pp. 833–843.

Melzack, R 2001, 'Pain and the neuromatrix in the brain', *Journal of Dental Education*, vol. 65, no. 12, pp. 1378–1382.

Mendelson, D 2002, 'Aspects of causation in hippocratic medicine and roman law of delict', in I Freckelton & D Mendelson (eds), *Causation in law and medicine*, Ashgate, Burlington, pp. 58–82.

Ricoeur, P 1984a, *Time and narrative volume 1*, trans. K McLaughlin & D Pellauer, University of Chicago Press, Chicago.

Ricoeur, P 1984b, *Time and narrative volume 2*, trans. K McLaughlin & D Pellauer, University of Chicago Press, Chicago.

Ricoeur, P 1992, *Oneself as another*, trans. K Blamey, University of Chicago Press, Chicago.

Ricoeur, P 2000, *The just*, trans. D Pellauer, University of Chicago Press, Chicago.

Rowe, P 1977, 'The demise of the thin skull rule?', *The Modern Law Review*, vol. 40, no. 4, pp. 377–388.

Safe Work Australia 2016, *The cost of work-related injury and illness for Australian employers, workers and the community: 2012–13*, Safe Work Australia, Canberra, viewed 15 January 2019, https://www.safeworkaustralia.gov.au/system/files/documents/1702/cost-of-work-related-injury-and-disease-2012-13.docx.pdf.

Stapleton, J 2002 'Scientific and legal approaches to causation', in I Freckelton & D Mendelson (eds), *Causation in law and medicine*, Ashgate, Burlington, pp. 14–37.

Tsalamandris, A & Lombard M 2011, 'Serious injury? Give credit where it's due', *Law Industry Journal*, vol. 85, no. 5, pp. 37–39.

Workplace Injury Rehabilitation Compensation Act 2013, viewed 15 January 2019, http://www.legislation.vic.gov.au/domino/web_notes/ldms/pubstatbook.nsf/f932b66241ecf1b7ca256e92000e23be/3629925065CDB2A6CA257C210015979B/$FILE/13-067a%20authorised.pdf.

CHAPTER 7

Racial Futurity: Biolegality and the Question of Black Life

Nadine Ehlers

Introduction

Assisted reproductive technologies (ARTs) are overwhelmingly celebrated as *techniques of affirmation*—that promote and enhance the capacity for the continuance of "life itself." Not only do they facilitate "traditional"/heteronormative family formation, but they have also been viewed as revolutionary because they invite—and indeed enable—"recombinant" donor-informed kinship ties that rework our very notions of "family" (Lenhardt 2017). At the same time, however, there is a cause for hesitation, as ARTs open new terrains for the *valuation* of life. We see this in a generalized sense in that such technologies can be called on to create "perfect" or "miracle" babies, with traits deemed desirable (more valuable) by prospective parents and/or society at large. In a more specific sense, they operate as means of the valuation of life or potential life, a fact particularly evident in terms of race. Not all lives are affirmed

N. Ehlers (✉)
Department of Sociology and Social Policy, University of Sydney,
Sydney, Australia
e-mail: Nadine.ehlers@sydney.edu.au

M. de Leeuw and S. van Wichelen (eds.),
Personhood in the Age of Biolegality, Biolegalities,
https://doi.org/10.1007/978-3-030-27848-9_7

equally in reproductive bioscientific-technological interventions: futurity is subject to the caesura of race.

Here I focus on the racialized and racializing dimensions of in vitro fertilization (IVF) in relation to a black/white racial divide as it operates in the United States, as it is this color-line that has been the most salient and enduring in what Hortense Spillers has called the "American grammar" of race: whereby "the ruling episteme that releases the dynamics of naming and valuation remains grounded in the originating metaphors of captivity and mutilation" (1987, p. 68). My focus is motivated by the continuing force that the black/white binarism yields and its ongoing currency and rehearsal in the contemporary U.S. racial landscape. In what follows, I explore the imbrications of IVF and race. My primary interest is to examine how IVF raises complex issues at the intersection of biology, race, and the law—specifically in terms of emerging forms of biolegality—through a focus on the 2014 Illinois Northern District Court case of *Cramblett v. Midwest Sperm Bank*.[1] The overarching concern is to ask: do IVF technologies represent new complexities for the denial of black life—where denial is understood as the negative valuation of black life?

Racial Valuation in IVF Technologies

To explore such a question, let me begin with a preliminary sketch of various ways that IVF technologies and practices can be said to promote race-based hierarchies and augment the differential distribution of racial valuation and vulnerability. Such operations are potentially pronounced as reproductive technologies become increasingly technologized and medicalized. This potentiality was noted already over two decades ago by Rayna Rapp and Faye Ginsburg, who argued that new bioscientific interventions into reproduction bolster a system of class-and race-stratified reproduction in which "some people are empowered to nurture and reproduce while others are disempowered" (1995, p. 3).[2] Indeed, it would seem that IVF promotes racial hierarchies precisely because they primarily support and optimize white life.

In the first instance, this promotion of racial hierarchies and white life in IVF can be seen in *factors of access to such technologies and the very design of the technologies themselves*, which many see as being geared toward white women and the specificities of their reproductive concerns (Quiroga 2007; Ginsburg and Rapp 1995; Murphy 2017).

For example, African American women are largely excluded from or disadvantaged within technologies of reproductive optimization, despite the fact that they have higher rates of infertility—at 11.5 percent (Armstrong and Plowden 2012, p. 651).[3] Middle to-upper-class white women access fertility treatment in the United States at more than twice the rate of African American (or Latina) women, who are more likely to have lower levels of education and household income compared to white women seeking infertility services (Daniels and Heidt-Forsythe 2012, p. 721). And, income becomes particularly important considering that the average cost for one IVF cycle in the United States is approximately $12,400, and many couples undergo multiple cycles before achieving a live birth, such that total treatment and medication costs can reach $85,000 (Humphries et al. 2016, p. 220). Moreover, the cost of a live birth is estimated at $41,132 (Armstrong and Plowden 2012, p. 652).[4] Racial disparities in IVF and ARTs more broadly are not only evident in the racial identity of users, however. Rather, as Camisha Russell has noted, "[w]e must consider how these technologies have arisen and developed with the concerns of a certain population in mind and what traces they bear of that particular standpoint" (2015, p. 604). Broad-based systemic infertility is not itself viewed as the primary problem. If it was, we could better address untreated Sexually Transmitted Infections (STIs) and poor medical treatment of women during an earlier birth, both of which cause infertility and disproportionately affect poor women of color; we would address environmental toxicity that leads to compromised fertility and poorer maternal health; and we would widen generalized access to preventative measures for infertility rather than focus on retroactive interventions. Instead, IVF is offered as seemingly the best way to address delayed childbearing: this is not a key cause of infertility among women of color but linked more to (largely white) female professionals and white-collar workers.

Second, the promotion of racial hierarchies in ARTs is evident in the *differential IVF outcomes based on race*. A recent report of ART in the United States noted that, of all the patient characteristics thought to influence ART success, "non-Hispanic black race" had the strongest negative association with a good perinatal outcome.[5] Thus, even if African American women do access ARTs, their outcomes are worse. They experience lower implantation rates and lower ongoing pregnancy rates, with black women having an ongoing pregnancy rate of only 14.9 percent against 38.8 percent for white women (Armstrong and Plowden

2012, p. 654). A single-site study of race-based IVF outcomes at Fertility Centers of Illinois found the following differences between black and white women: black women had fewer mature oocytes, a lower rate of fertilization of these oocytes, fewer five-day transfers of fertilized oocytes, a lower surplus of blastocysts (to cryopreserve for future transfers), and higher cancelation rates of cycles (for example, failed cycles) compared to white women. They also found that black women's clinical pregnancy rate was lower than white women's (at 24.4 percent compared to 36.2 percent), they had higher rates of spontaneous abortion (at 28.9 percent compared to 14.6 percent), and worse oocyte quality overall (McQueen et al. 2015).[6] Numerous factors could account for these discrepancies and the general obstacles to pregnancy through IVF that African American women face. Among these are increased infertility due to environmental factors (exposure to toxins), smoking, drug use, higher rates of obesity, and general exposure to ongoing and systemic discriminatory healthcare practices (Armstrong and Plowden 2012).

Third, a more specific promotion of racial hierarchies is at work in the *organization of donor sperm services.* Donor services reinforce racial delineation in the simple fact that they separate their directories by race—where semen is racially classified. For example, until recently, California Cryobank stored and shipped semen in color-coded vials (black, white, yellow, and red [for mixed-raced donors]). Most importantly, such a practice ontologizes race: it augments beliefs that race is biological, that it is inheritable, and that it is essential to the individual—ideas have been widely critiqued and refuted.[7] Yet, while we might understand race-based categorization of sperm as a simple reflection of consumer demand—for phenotypical or supposed genotypical sameness—donor services must also be seen as *accentuating* race in ways that invite individuals to exclude *all* donors of a particular race. This secures the idea that race is paramount in family formation and primes individuals to always-already think in racial terms (Fox 2009). More than merely reproducing racial classification, however, donor services reinforce the negative relation of blackness: they often push "elite sperm" from men who have attended four-year colleges, with the suggestion that the highest quality sperm comes from these men, thus reinforcing stratified human value, where lower income men (often minorities) are at the bottom. The donor catalogue is thus a proxy for racial pedigree (Quiroga 2007, p. 150). Through both these factors, donor services endorse a white kinship model—a breeding schema—that works to promote whiteness and maintain fears about racial mixing, particularly between whites and non-whites.[8]

In these various ways, ARTs and the donor sperm industry reinforce racial delineation, the negative relation of blackness, and the valuing of white life and reproductive futurity. As I claimed in the opening of this piece, such technologies and practices also present new possibilities for the *valuation* and *adjudication* of life before the law.[9] These technologies might enhance the capacity for the continuance of "life itself," but law can (and has) often become the site of arbitration for disputes related to the use of these technologies—around what counts as life, what lives should be enhanced or enabled, and whether a life is eligible for life as such (Dillon and Neal 2008, p. 168).

The Caesura of Race in Cramblett v. Midwest Sperm Bank

The case of *Cramblett v. Midwest Sperm Bank* raises questions along these lines, but specifically through the prism of race—making it a salient example of the imbrications of race and IVF and the caesura of race in reproductive futurity. In this case, Jennifer Cramblett, a 36-year-old woman from Uniontown, Ohio, filed suit against Chicago-area Midwest Sperm Bank because the clinic artificially inseminated her with the "wrong sperm." Cramblett claimed she and her partner Amanda Zinkon—both white—chose a "blond haired, blue-eyed individual" so that the child would closely resemble her partner (Cuevas 2014). Due to an administrative error, they instead received the sperm of a black donor, a fact Cramblett only found out approximately five months into the pregnancy. In her lawsuit, Cramblett claimed she lives with "fears, anxieties and uncertainty" about her and the child's future in their all-white racially intolerant town, and in her own family, which she claimed lacks cultural competence concerning African Americans (*Cramblett v. Midwest Sperm Bank LLC* 2014, s. 23).[10] Cramblett filed what is referred to as a *wrongful birth* suit and sought more than $50,000 damages for "personal injuries, medical expenses, pain, suffering, emotional distress and other economic and non-economic issues" (*Cramblett*, s. 31).[11] The Complaint argued that these factors were not only present concerns but would continue to resonate "in the future" (*Cramblett*, s. 31).

How, then, can the birth of a black child—to a white mother—be conceived as a "wrongful birth"? In this assessment of biolegality, legitimate birth, and indeed legitimate personhood, *Cramblett* squarely placed the valuation of black life before the law, highlighting what Dorothy Roberts identifies as the "potential horror" involved in reproductive technologies (2009, p. 787). What does this case rest on? What new complexities does it represent?

Ordinarily, the charge of "wrongful birth" is a claim brought by the parent(s) of a child born with a birth defect or disability against medical personnel, whose negligence in detecting or disclosing said defect deprived the parent(s) of the opportunity to make an informed decision about whether to avoid or terminate the pregnancy. It is a claim of a civil "wrong" on the basis of the ***denial of the right to avoid the birth of a child with a (said) defect***. And, the denial of this opportunity in wrongful birth tort is framed as an ***injury***. Both medical counseling around termination of fetuses with disabilities and subsequent wrongful birth tort effectively, then, enter into what Michelle Kelsey Kearl characterizes as a biopolitics that normalizes termination: they constitute "a biopolitics around which characteristics should be reproduced and which should be prohibited in a population" (2018, p. 202). Indeed, here we see an instantiation of the biopolitical logic that certain forms of life are a biological ***threat*** to be defended against.[12] In such cases, what is being asked of the court is a recognition of injury to the parent(s), generally through pecuniary damages, to financially compensate for bearing the burden of this life. For wrongful birth claims, the actual being-in-the-world of the 'defective child' is a supposed legally cognizable injury.

Clearly, much can be said about the compulsory ableism that wrongful birth tort relies on and augments. Briefly, this ableism positions only certain bodies and capacities as normative and devalues those seen as deviating from these regularized standards. It operates within and depends on a medical model that naturalizes disability as biological deficiency and abnormality. Moreover, "ableism flourishes within capitalism by devaluing any people who cannot conform to expectations for productivity … and within ableist society and the pathology paradigm, disabled existence is considered unworthy, lacking in value, and undesirable" (Brown 2018, n.p.). Rather than unpacking these logics here, I want to travel in a different direction.

In the *Cramblett* case, there was no defect, per se. There was no suggestion that the child had a physical or intellectual disability or any serious medical condition. There was ***no disabled subject***. Yet the claim of wrongful birth in this case has internal to it the notion that the child should not have been born—that she was "wrongfully born." What is wrong, here, is the child's race: that the child is black. And, thus, blackness is situated as synonymous with defect and disability. Perversely, however, the supposed disabled subject disappears in this case and, as Kearl has noted, "is replaced by a suffering able-bodied white woman" (2018, p. 307). The mother's suffering is positioned as being caused by the child, Payton, and it is the child who is framed as posing an inherent harm. Indeed, blackness is positioned *as injury* (to the mother, to whiteness) inaugurated through the bioscientific-technological intervention of IVF donor semination.

Importantly, race can only be used as a measure of injury in the context of a generalized negative valuation of black life and within an epistemological framework that presupposes that black negativity can incur on white life. I will address these two presuppositions through exploring two intersecting points of suffering that this child supposedly produces, as suggested by the legal Complaint.

The first form of suffering ostensibly caused by the child is *a perceived loss over genetic parentage*. As Cramblett noted in her Complaint, the couple's "desire was to find a donor [and have a child] with genetic traits similar to both of them" (*Cramblett*, s. 9). This loss over perceived genetic parentage links in part with the idea that IVF can be a means by which to create "perfect babies." IVF generally involves selecting the healthiest embryos, with characteristics valued by the user/s and broader society. Genetic testing during the first trimester has become so routine, particularly in IVF or artificial insemination pregnancies, that the demand for perfect babies has become almost thoroughly naturalized. And reprogenetics—or germ-line engineering—goes a step further to edit the DNA of reproductive cells or the embryo itself, with the promise of correcting disease genes and passing those genetic fixes on to future generations.[13] Donor IVF, such as that used by Cramblett, enables individuals to choose donor's that possess particular heritable and non-heritable characteristics, predicated on disease profile, body type, education level, family behavioral/mental health background, religion, and race (among other features) (Kearl 2018, p. 302).

Part of this selecting a donor is based on producing a form of relatedness that Laura Mamo calls "affinity-ties." She describes this production as "a kinship device co-constructed between … [clients] imagining shared social and cultural characteristics and sperm banks' investment of the biomaterial, sperm, with an ability to create relatedness"

(2005, pp. 237–238). Relatedness—the affinity-tie—is achieved by picking sperm that will have an affinity to the IVF user both in a socio-cultural and loose biogenetic sense. Such affinities can be particularly important when users of reproductive technologies engage donors to create "family." In one sense, then, technology enables "a transformation of the markers of kinship, family, and relatedness" (Mamo 2005, p. 238), as it can be used to challenge the notion that kinship is based on biology—calling instead on a donor to facilitate kinship. Simultaneously, however, it can also be used to reaffirm the importance of biogenetic ties: by choosing a donor who resembles the legal parent(s), users reassert the importance of supposed biogenetic affinity.

Recall, Cramblett desired to have a child that was 'similar' to both herself and her partner. In her Complaint, she specifically states she wanted the potential child to have similar genetic traits. We can understand this to mean that Gramblett and her partner desired that the child would *look* like them: clearly, this would mean that the potential child would racially resemble the couple.[14] This sentiment was clearly stated in the Complaint, where Cramblett outlined how she first learnt of the "mix up": the receptionist at the sperm bank asked if she had requested an African American donor, to which Cramblett replied, "No, why would I request *that*? My partner and I are Caucasian" (*Cramblett*, s. 15, emphasis added). The affinity-tie being created, then, was one of biological and familial whiteness. However, the birth of a child socially defined as black foreclosed this possibility, and Cramblett's suffering can in part thus be understood as a loss of (white) lineage. This loss is only possible due to the fact that within the American racial lexicon whiteness can only exist in supposed "pure" "undiluted" form: it must be unambiguous whiteness. The entire history of U.S. anti-miscegenation jurisprudence and laws pertaining to racial classification of offspring from black/white intimacy attests to this.[15] The perceived loss is also informed by a eugenic inheritance. As Suzanne Lenon and Danielle Peers have noted:

> It is within this neo-eugenic context that the birth of a 'non-white' child to a 'white family' through neo-eugenic reproductive technologies can be understood as such an unthinkable harm, and further, can be legally pursued through legal action designed to support parental rights to prevent disability, defect, and (racial) degeneracy. (2017, p. 154)

Blackness can only exist as a harm precisely because it has been cast in the post-Enlightenment Western imaginary as deviant and inferior to normative whiteness. The idea that blackness represents negativity has a torturous history. We know that within the Western racial lexicon, if the Human subject is predicated on social recognition, volition, subjecthood, and the valuation of life, blacks were exiled from this relation—as negative *absence* (Hartman 1997, p. 21).[16] These ideas live on in the afterlife of slavery and structure our present. The harm, however, as it appears in the context of this case, is that this negativity has been inflicted on a white woman. Again, within Western racial logics, white women have been charged with the responsibility of bearing white children and, through this, maintaining a possessive investment in whiteness (Zack 2000). This opportunity is lost to Cramblett due to the very technological intervention that was meant to facilitate racial sameness. As Saidiya Hartman (2014) has noted, "Payton dispossesses her mother by being born, taking the space of a more qualified, more desired white candidate, erupting into the world as damaged goods." If the first point of Cramblett's suffering is a perceived loss over genetic parentage, then we might understand this as a mother having lost the property of her (imagined) child's body and the future potential capital (social and otherwise) that body/subject might yield.

The second point of suffering caused by the child is that she inaugurates the mother's loss of white property (in herself). Here I refer to legal scholar Cheryl Harris' seminal work proposing that U.S. law created whiteness as a desirable and beneficial a form of property. Whiteness has operated as a form of property in the following ways: it has included the *rights of use and enjoyment*—it can be used as a deployable resource at the social, political, and institutional level to maintain control and privilege; it is a *type of status property*—people have a reputational interest in being white; and it *involves the right to exclude* (Harris 1993). The child causes injury—indeed she *is* injury—because her entry into the world causes the mother to lose her sense of white privilege: the child depreciates the mother's white property-value. Because of the child, Cramblett now has to travel into non-white spaces. As she outlines in her Complaint (s. 24), the very act of having to attend to the child's hair means that she needs to visit black hairdressers in black neighborhoods. As Lenon and Peers (2017, p. 158) have argued, "[t]his forced mobility disrupts the 'settled expectations' (Harris 1993) of middle-class [white] convenience to resources and services within segregated white spaces." This need to travel also unsettles her race-based comfort. She states that she feels both hypervisible and "not overtly welcome" in these spaces (*Cramblett*, s. 24). For white people, who generally move through the world unencumbered and invisible in their normativity, these new proximities appear to cause Cramblett harm. Lastly, the child opens Cramblett to injury because she will now be exposed daily to the racism of both her wider community and immediate family. In the Complaint, Cramblett states that "she was raised around stereotypical attitudes about people other than those in her all-white environment" and that "[f]amily members, one uncle in particular, speaks openly and derisively about persons of color" (*Cramblett*, s. 22). What ostensibly underscores Cramblett's grief, then, is that she will no longer be unaffected by the racism around her—but bound up in it (and subject to it) due to her proximity (relation) to the child. She can no longer live comfortably in this racist environment. Rather than presenting racism as the harm, then, the child is the harm that evacuates Cramblett's ability to freely "use and enjoy" whiteness.

Conclusion: Procreation's Racial Order

The Illinois District Court ultimately dismissed the case, with the judge ruling that the suit could not proceed under a "wrongful birth" claim.[17] On the one hand, we could view this as a legal refusal that this was a wrongful birth and a seeming legitimation of black racial life and value. We could read the court as stipulating the biolegitimacy of this child. This is not what happened. Instead, the motion to dismiss emphasized that legal standards do not recognize claims for healthy babies and only disability is the rightful subject of wrongful birth claims. In doing so, the court simultaneously upheld dominant notions of disability (as pathological and potentially wrongful) and *failed to address race at all.*[18] By this I mean that no mention was made of race in the dismissal and, as such, the issue that was at the very heart of the wrongful birth claim was silenced and rendered invisible. At a minimum, the judge could have made the claim that black lives matter. Instead, we must view the dismissal (on the terms that it was made) as the abdication of legal responsibility to address/redress enduring racism *and the very conditions of possibility that gave rise to this case in the first place.*

I began this piece with the claim that ARTs in general, and IVF in particular, reproduce race-based hierarchies and work to affirm white life. More than this, however, my central question was to ask whether IVF technologies open new terrains and complexities for the denial of black life: that is, the racialized *valuation* of life and the adjudication of life before the law. What we see in this case, however, is that what is inaugurated are not new complexities at all. *Cramblett* shows us that donor IVF simply opens the door for rearticulating the terms of reproduction through the lens of centuries-old anti-miscegenation rhetoric: that whites and blacks should not mix. *Cramblett* shows that donor IVF can be the very means through which to reproduce dominant knowledges of race and race itself. This case highlights that the symbolism of free-market choice in new reproductive technologies is clearly not just about producing healthy babies. It would seem instead that these technologies are being called on to maintain "a racial calculus and a political arithmetic that were entrenched centuries ago" (Hartman 2007, p. 6). As much as these technologies open the possibilities for new forms of kinship and co-belonging, they re-secure the reality that kinship is seen to have a color. Moreover, these technologies can be used to cement the differential valuation of life according to racial hierarchy (choosing white children over other kinds of children) and to reproduce the negative relation that marks black existence (by choosing *against* the possibility of having a mixed child). Ultimately, then, "[r]egardless of how technology has freed us … [w]e still have the messy nucleus of procreation's racial order to contend with" (Holland 2012, pp. 61–62)—a racial order that in many ways persistently secures white life in and against black life and forecloses, either literally or discursively, black futurity.

Notes

1. *Cramblett v. Midwest Sperm Bank, LLC.* 'Complaint for Wrongful Birth and Breach of Warranty.' 29 September 2014. Circuit Court of Cooks County, Illinois. Also see the motion to dismiss, *Cramblett v. Midwest Sperm Bank, LLC. No. 15 L 000282.* Note that Cramblett filed her original complaint in Cook County Circuit Court, and the case was transferred to DuPage County on March 24, 2015. The complaint alleged *both* wrongful birth and breach of warranty pursuant to the Illinois Blood and Organ Transaction Liability Act (745 ILCS 40/3).
2. Also see Quiroga, who notes: "[a]s the U.S. infertility industry is fond of saying, their aim is to 'create families.' What remains unspoken is the desire to create a certain type of family, one that closely matches, and thus reproduces, the heteropatriarchal model of a white nuclear family" (2007, p. 144).
3. For 2017 statistics, see Martin et al. (2018).

4. Note, these technologies are rarely publicly funded, but instead require individual private resources. See Nanette R. Elster (2005), who argues that financial barriers are not the only factor restricting African American women's access to ARTs.
5. See Humphries et al. (2016, p. 212, n12), who state: "[l]arge, controlled studies that assess the effects of race/ethnicity on IVF outcomes are relatively sparse. The SART data from 2004 through 2006 provide the strongest evidence of disparities for black, Asian, and Hispanic women, and these data are bolstered by similar findings in 1999 through 2000 for black and Asian women."
6. Note, this study accounted for financial access, as more than 80 percent of participants had at least partial insurance coverage (access to care is increased by state mandate in Illinois).
7. For a representative but by no means exhaustive sample, see Ian F. Haney López (1996), Michael Omi and Howard Winant (1986), Tukufu Zuberi (2001), David Theo Goldberg (1992), Henry Louis Gates (1987), and Troy Duster (2005).
8. On the history of this breeding schema, see Naomi Zack (2000).
9. Needless to say, intersections with the economy can also not be dispensed with. See Erin Heidt-Forsythe (2017).
10. Hereafter cited as *Cramblett.*
11. Note that the civil action was for both wrongful birth and breach of warranty. I focus on the wrongful birth aspect of the Complaint here.
12. See Michel Foucault (2003).
13. See Knowles and Kaebnick (2007). Also see Regalado (2015).
14. Here I am suspending discussion around the fact that race is not genetic. For work that does address this see, for instance, Alondra Nelson (2016) and Catherine Bliss (2012).
15. See Nadine Ehlers (2011).
16. Such ideas can be placed within a theorization of anti-blackness, understood not as racist acts or violence against the black body, but as "the violence that *positions* sentient beings outside the realm of the Human" (Bliss 2015, p. 89). On anti-blackness in relation to the *Cramblett* case, see Savannah Shange (2014).
17. See both the Motion to Dismiss and Clifford Ward (2015). Note that the case was dismissed on September 3, 2015 on two fronts: not only the wrongful birth claim but also the claim for breach of warranty. The judge concluded that 'breach of warranty' under a state law only covered blood and tissue donations, and that sperm was not tissue. Thus, the claim of breach of warranty did not stand. Cramblett's charges against Midwest Sperm Bank continued in various forms however. See *Cramblett v. Midwest Sperm Bank* 2017 IL App (2d) 160694-U.

18. A similar argument can be made for disability. Investigating that question, however, alongside its intersections with race, would be a separate project. For an analysis of the problematic of wrongful birth tort for disability, see Lydia X.Z. Brown (2018). Simply saying that Cramblett could have alternatively leveraged torts of fraud or medical malpractice (taking aim at the clinic) still keeps in place the idea that the 'black child' should not have been born.

Bibliography

Armstrong A & Plowden, TC 2012, 'Ethnicity and assisted reproductive technologies', *Clinical Practice (London)*, vol. 9, no. 6, pp. 651–658.

Bliss, C 2012, *Race decoded: the genomic fight for social justice*, Stanford University Press, Stanford.

Bliss, J 2015, 'Hope against hope: queer negativity, black feminist theorizing, and reproduction without futurity', *Mosaic: A Journal for the Interdisciplinary Study of Literature*, vol. 48, no. 1, pp. 83–98.

Brown, LXZ 2018, 'Legal ableism, interrupted: developing tort law and policy alternatives to wrongful birth and wrongful life claims' *Disability Studies Quarterly*, vol. 38, no. 2, n.p.

Cramblett v. Midwest Sperm Bank, LLC, 'Complaint for Wrongful Birth and Breach of Warranty', 29 September 2014. Circuit Court of Cooks County, Illinois.

Cramblett v. Midwest Sperm Bank, LLC. No. 15 L 000282, 'Midwest Sperm Bank LLC's Motion to Dismiss Plaintiff's Complaint at Law', 2015. Circuit Court of Cooks County, Illinois.

Cuevas, M 2014, 'Ohio woman sues sperm bank after racial mix-up', CNN, 2 October, viewed 15 November 2018, https://edition.cnn.com/2014/10/02/us/sperm-bank-race-lawsuit/index.html.

Daniels, CR & Heidt-Forsythe, E 2012, 'Gendered eugenics and the problematic of free market reproductive technologies: sperm and egg donation in the United States', *Signs: Journal of Women in Culture and Society*, vol. 12, no. 3, pp. 719–747.

Dillon, M & Neal, AW 2008, *Foucault on politics, security and war*, Palgrave Macmillan, Hampshire.

Duster, T 2005, 'Medicine, race and reification in science', *Science (American Association for the Advancement of Science)*, vol. 307, pp. 1050–1051.

Ehlers, N 2011, 'Onerous passions: colonial anti-miscegenation rhetoric and the history of sexuality', *Patterns of Prejudice*, vol. 45, no. 4, pp. 319–340.

Elster, NR 2005, 'ART for the masses? Racial and ethnic inequality in assisted reproductive technologies', *DePaul Journal of Health Care Law*, vol. 9, no. 1, pp. 719–733.

Foucault, M 2003, *Society must be defended: lectures at the Collège de France 1975–1976*, Picador, New York.

Fox, D 2009, 'Racial classification in assisted reproduction', *The Yale Law Journal*, vol. 188, pp. 1844–1898.

Gates, HL 1987, *Figures in black: words, signs, and the 'racial' self*, Oxford University Press, New York and London.

Ginsburg, FD & Rapp, R 1995, *Conceiving the new world order: the global politics of reproduction*, University of California Press, Berkeley.

Goldberg, DT 1992, 'The semantics of race', *Ethnic and Racial Studies*, vol. 15, pp. 543–569.

Harris, C 1993, 'Whiteness as property', *Harvard Law Review*, vol. 106, no. 8, pp. 1707–1791.

Hartman, S 1997, *Scenes of subjection: terror, slavery, and self-making in nineteenth-century America*, Oxford University Press, New York.

Hartman, S 2007, *Lose your mother: a journey along the Atlantic slave route*, Farrar, Straus, and Giroux, New York.

Hartman, S 2014, 'The value of whiteness: a lawsuit is being waged against the "wrongful birth" of a black child', *The Nation*, 12 November, viewed 11 August 2018, https://www.thenation.com/article/value-whiteness/.

Heidt-Forsythe, E 2017, 'Morals or markets? Regulating assisted reproductive technologies as morality or economic policies in the states', *AJOB Empirical Bioethics*, vol. 8, no. 1, pp. 58–67.

Holland, SP 2012, *The erotic life of racism*, Duke University Press, Durham and London.

Humphries, LA, Change, O, Humm, K, Sakkas, D & Hacker, M 2016, 'Influence of race and ethnicity on in vitro fertilization outcomes: systematic review', *American Journal of Obstetrics & Gynecology*, vol. 24, no. 2, pp. 212–224.

Kearl, MK 2018, 'The stolen property of whiteness: a case study in critical intersectional rhetorics of race and disability', *Rhetoric Review*, vol. 37, no. 3, pp. 300–313.

Knowles, LP & Kaebnick, GE 2007, *Reprogenetics: law, policy, and ethical issues*, Johns Hopkins University Press, Baltimore.

Lenhardt, RA 2017, 'The color of kinship', *Lowa Law Review*, vol. 102, pp. 2071–2107.

Lenon, S & Peers, D 2017, ''Wrongful' inheritance: race, disability and sexuality', *Cramblett v. Midwest Sperm Bank*, *Feminist Legal Studies*, vol. 25, pp. 141–163.

López, IH 1996, *White by law: the legal construction of race*, New York University Press, New York and London.

Martin, JA, Hamilton, BE, Osterman, MJK, Driscoll, AK & Drake, P 2018, 'Births: final data for 2017', *National Vital Statistics Reports*, vol. 67, no. 8.

National Center for Health Statistics: Hyattsville, MD, viewed 15 November 2018, https://www.cdc.gov/nchs/data/nvsr/nvsr67/nvsr67_08-508.pdf.

Mamo, L 2005, 'Biomedicalizing kinship: sperm banks and the creation of affinity-ties', *Science as Culture*, vol. 14, no. 3, pp. 237–264.

McQueen, DB, Schufreider, A, Lee, SM, Feinberg, EC & Uhler, ML 2015, 'Racial disparities in in vitro fertilization outcomes', *Fertility and Sterility*, vol. 104, no. 2, pp. 398–402.

Murphy, M 2017, *The economization of life*, Duke University Press Books, Durham, NC

Nelson, A 2016, *The social life of DNA: race, reparations, and reconciliation after the genome*, Beacon Press, Boston.

Omi, M & Winant, H 1986, *Racial formation in the United States: from the 1960s to the 1980s*, Routledge, New York and London.

Quiroga, SS 2007, 'Blood is thicker than water: policing donor insemination and the reproduction of whiteness', *Hypatia*, vol. 22, no. 2, pp 143–161.

Regalado, A 2015, 'Engineering the perfect baby', *MIT Technology Review*, May/June, viewed 11 November 2018, https://www.technologyreview.com/s/535661/engineering-the-perfect-baby/.

Roberts, D 2009, 'Race, gender, and genetic technologies: a new reproductive dystopia?', *Signs: Journal of Women in Culture and Society*, vol. 34, no. 4, pp. 783–804.

Russell, C 2015, 'The race idea in reproductive technologies: beyond epistemic scientism and technological mastery', *Journal of Bioethical Inquiry*, vol. 12, pp. 601–612.

Shange, S 2014, 'Black on purpose: race, inheritance and queer reproduction', *The Feminist Wire*, 10 October, viewed 1 November 2013, http://thefeministwire.com/2014/10/black-purpose-race-inheritance-queer-reproduction/.

Spillers, H 1987, 'Mama's baby, Papa's maybe: an American grammar book', *Diacritics*, vol. 17, no. 2, pp. 65–81.

Ward, Clifford 2015, 'Suit filed over mix-up at Downers Grove sperm bank is dismissed', *Chicago Tribune*, 3 September, viewed 15 November 2018, https://www.chicagotribune.com/suburbs/downers-grove/news/ct-dupage-sperm-bank-suit-met-0904-20150903-story.html.

Zack, N (ed.) 2000, *Race/sex: their sameness, difference and interplay*, Routledge, New York.

Zuberi, T 2001, *Thicker than blood: how racial statistics lie*, University of Minnesota Press, Minneapolis.

PART III

Governing Persons

CHAPTER 8

Phenotypic Personhood: Epigenetics and the Biolegality of Processing Asylum

Zsuzsanna Dominika Ihar

Introduction

Over the last decade epigenetics has been heralded for its capacity to articulate a new lexicon of the body and personhood; one that has the ability to acknowledge not only the influence of environmental factors on one's development—for example, the intergenerational markings of trauma, neglect, and poverty—but also the historical and political forces that shape us in specific and situated ways (Mayr 1980). It is a lexicon whose native speakers include molecular biologists, geneticists, and various actors in the bio and social sciences (Niewöhner 2011; McGowan and Szyf 2010). In recent years, however, a distinct shift has occurred in its uptake and articulation. Agencies and actors in border processing and control have demonstrated a marked keenness for the language of epigenetics, citing its use in generating a molecularized version of the migrant

Z. D. Ihar (✉)
Department of Sociology and Social Policy, University of Sydney,
Sydney, NSW, Australia
e-mail: Zsuzsanna.ihar@sydney.edu.au

M. de Leeuw and S. van Wichelen (eds.),
Personhood in the Age of Biolegality, Biolegalities,
https://doi.org/10.1007/978-3-030-27848-9_8

subject—a version that is trackable, consistent, and streamlined. This chapter examines the manifestations of epigenetic and DNA data within these border control agencies, with specific attention directed toward bio-genomic verifications of age, nationality, and health. It will do so in order to agitate for the development of adequate legal infrastructures, providing forms of advocacy and protection for individuals most vulnerable when it comes to these bio-technical methods.

Reflecting on forms of politicized epigenetic storytelling within the fields of literature and the environmental humanities, the first part of this chapter will focus on the alternate uptake and narrativization of epigenetics within the emergent site of border science. It will hone in on the particular transformation of biological populations into lucrative sources of mineable data, utilized by both corporations and governments for the purposes of streamlined identification, processing, as well as management. The section will also tap into a broader history of interpretation when it comes to genomic data, depicting it as something fixed, collectible, and commodifiable. Biology in this context will be seen as not only capable of generating putative risk loci and health profiles, but also forms of legal personhood and techniques of evidence-making.

Part II will provide an example of bioscientific heuristics sliding into the realm of legal cross-checking, tilting its gaze at the willful configuration of epigenetics as a border apparatus. Introducing the Hildesheim case—the first instance where an epigenetic clock was used to confirm/contest a claimant's alleged age by authorities—the section will examine the stakes of measuring rates of maturation outside of the matrices of transmissive trauma or genetic injury. It will argue that the technology essentially created a contained molecular profile, which ignored the aging impacts of war, border control, detention, and other psycho-physiological stresses. In doing this, age was effectively localized within the internal environment of the genetic.

Part III of this chapter will provide another contemporary example of epigenetic border science. It will examine the use of mitochondrial DNA (mDNA) and Y chromosome testing by the United Kingdom (UK) Border Agency (Home Office 2010), as well as their focus on single-nucleotide polymorphisms (SNPs), in a project intended to verify migrant claims of nationality. Continuing the argument of epigenetics binding the identity of asylum seekers to reductive forms of intelligibility, I will focus on the use of epigenetics as a novel territory of securitization, allowing indeterminate biophysical markers to be re-conceptualized as stable and tethered to the subjects of border processing.

Finally, I will introduce the concept of "phenotypic personhood," where phenotypical responses to environmental, ecological, historical, and social conditions have been reformulated into injurious categories of criminality and suspected deceitfulness.

The Movement from Biometrics to Biomarkers in Border Science

There are modes of storytelling enmeshed in the body; macroscopic and microscopic indentations, microbial citations, epigenomic anecdotes, all coming together to tell complex narratives about the person they collectively compose. In the words of Donna Haraway, these biological tidbits exist "in partial and awed translations across difference" (Haraway 2016, p. 10), providing open windows to cultivate storied practices—practices which are capable of shaping bodily matter in new and uncanny ways.

The epigenetic, particularly in the context of the postgenomic age seems to possess narrative expandability, allowing multiple actors and relations to be recognized, complicating clear cut lines around biology. It is a field defined by unstable categories and dynamic processes like methylation, the presence noncoding Ribonucleic Acids (RNAs), previously seen as "junk," and histone modifications, which evade the determinism of a permanently altered DNA sequence. The discipline of epigenetics provides a new epistemological terrain for storytellers working across the sciences and the humanities. New narratives are told about how "toxins, work stress, nutrition, socio-economic status, maternal care" (Meloni 2016, p. 3)—as well as the ghostlier matter of inter- and trans-generational trauma—materialize in phenotypic form. Indeed, findings in the field of epigenetics have become essential intertextual citations in wider narratives of traumatization and socio-political violence. We see, for example, inherited Post Traumatic Stress Disorder (PTSD) and low cortisol levels among the children of Holocaust survivors (Glausiusz 2014), where epigenetics as a framework has allowed for the firm re-centering of causality within generations who otherwise seem removed from the original historical context (and time frame) of genocidal violence; to the impact of receding social support systems on families caught in cycles of poverty and state neglect. Perhaps it is this particular potentiality, one of carving out a different chrono politics of consequence, of traveling beyond strictly set time-frames and neat categorizations of harm, which make the appearance of an epigenetic lexicon within matters of asylum and migration so evocative and potentially generative (Lock 2015).

Indeed, leafing through pages of *The Displaced: Refugee Writers on Refugee Lives* (Nguyen 2018) on a nondescript weekday, with my eyes eventually settling on Maaza Mengiste's chapter (titled This Is What the Journey Does), the epigenetic began to materialize in salient, literary form—giving a hint of what genomic modes of storytelling could look like. Describing the seemingly erratic actions of a young East African man of Eritrean, perhaps Ethiopian, origin, who freezes in the middle of an intersection only to wildly wave his arms, moving waywardly, Mengiste (2018), flatly yet movingly attests to the fact that the man did "not leave home like this" (p. 70). Rather, the man's inscribed impulses, like pockmarks and gashes left by methods of torture, are the bodily traces and remnants of "what journey does" (p. 70). It is a moment where the metaphoric gestures in Mengiste's work seem to coalesce into a newly recognizable avenue for epigenetic research, where the genomic is reconfigured into an *active* method by which seemingly unintelligible pain and violence can be translated and rendered "sensible" (Lock 2015, p. 161).

However, in recent scholarship around migratory processing there has been a movement away from the development of a politicized form of scientific storytelling, with the interaction between epigenetics and migration skewing toward a purely functional and administrative uptake (Vukov 2016). This chapter interrogates this particular skew, highlighting the strategic incorporation of epigenetic findings for the bolstering of verification processes, the streamlining of refugee background checks, and the elaboration of methods which clump migrant/refugee applicants into imaginary, and often arbitrary, categories of identity and personhood (Maguire and Rao 2018). The increasing prevalence of epigenetic profiles being utilized by border authorities to cross-check claims and re-prioritize the body as the most accurate site of documentation (in opposition to paperwork and anecdotal accounts) echoes Ruth Benjamin's concept of "strategic calibration" (2009, p. 343), where genomic data is mobilized to "match the socio-historical record and a re-imagining of historical and cultural narratives to make sense of genomic findings" (p. 342). In this way, the state's interest in bio-genomic data moves away from being one of convenience, with biomarkers functioning merely as another form of information and evidence, to an active exercise in the transformation of the genomic into a valuable means of "storying" the body strategically (Haraway 2016, p. 39).

This indeed funnels into a wider trend within the sphere of genomics, where the biology of the population "becomes a natural resource" (Benjamin 2009, p. 344)—one that assists with nation-building projects, the development of profitable pharmacogenomic markets (*biotechnic markets*), the securitization of "biopolitically rogue populations" (Benjamin 2009, p. 343), and the creation of profit-driven biobanks. While this cross-pollination between state institutions and corporate genomic industries has hitherto occurred indirectly—one may be reminded of the use of DNA testing and ancestry websites (including FamilyTreeDNA) by Canadian immigration officials to confirm nationality (IANS 2018)—here is a sense that noncommercial interactions can serve as a blueprint for the transformation of migrant bodies into tradable and traceable bio-commodities.

From this we can see that both the genomic and the epigenetic are invested with an enormous potential to become sites where "biotech start-ups, spin-offs, and subsidiaries" (Thacker 2005, p. xv) entangle with governmental bodies, dovetailing developments in the biosciences with national security concerns and the desire to cultivate lucrative databases. This was the case when congresswoman Jackie Speier called for the involvement of 23andMe, a private DNA-testing company, in supplying kits to families separated at the United States (US)-Mexico border with the hope of securing their potential reunification (Zhang 2018). Within days, Anne Wojcicki, the co-founder and chief executive officer of the company, assured her support for the venture and offered to donate kits and resources to complete large-scale genetic testing (Ho 2018). Another US-based DNA-testing company, MyHeritage, also committed publicly to formalizing this project, even announcing their intention to incorporate relevant government agencies to facilitate its logistics (Molteni 2018). These pronouncements framed the contours of what can be considered "on-the-spot" bioethics around the issue of DNA testing for the purpose of establishing kinship, fostering a sense of the project's permissibility while effectively effacing a thorny historical lineage of genomic exploitation and the violence of blood kin itself (Salter 2017; La Spina 2017). Furthermore, it has served as an invitation for

US Department of Health and Human Services to begin conducting its own DNA tests (Genome Web 2018), normalizing the use of commonly available forms of DNA testing for the purpose of verifying claimed relationships, prior to reuniting potential family members (Zhang 2018).

Ultimately, the keenness of the state and the corporate to invest in genomic, as well as epigenetic, methods of knowing—creating lucrative methods of managing migration and streamlining refugee claims—aligns with Browne's prediction of the advent of digital epidermalization (2009, p. 132). We are increasingly seeing bodies, which are deemed suspicious insofar that they exist outside of the protective jurisdictions of citizenship, become "worked wholesale through databases" (2009, p. 136); databases which are capable of clustering "individuals into certain segments…with the purposes of identifying risks, potential markets, or both" (p. 136). This not only reflects the expanding sphere of the technological regulation of bodies, moving away from a purely physical policing of territorial lines, but also the (re)introduction of forms of state regulation that are specifically bioscientific in nature (Bohmer and Shuman 2018), echoing Deleuze's prediction of a shift from societies of enclosure to societies of control (Deleuze 1992). The next section will attest to the implications of the postgenomic age for border technologies, urging us to recognize the fact that shifts in genomic paradigms and frameworks are not containable within zones of the laboratory (Landecker and Panofsky 2013): they frequently overflow, shaping, and contouring socio-political landscapes. Indeed, the *enfolding* of a developing (Moore 2015), or reactive, genome within the biosciences—"subject to time and space, biography and milieu" (Meloni 2016, p. 3)—*unfolds* into an active tool for border authorities, with novel affordances when it comes to the controlling and orientating the migrant subject (Jones 2016). It can be argued that such a technique feeds into the advent of the bio-legal within border science, where there is an ever-tighter link between bio-technical apparatuses and legal claim-making.

Bio-polygraphs and the Suspicious Body of the Unaccompanied Minor

Echoing Margulis and Sagan's conceptualization of life as always "preserving the past, making a difference between past and present" (2000, p. 86)—that is, of life being capable of "binding time, expanding complexity and creating new problems for itself" (p. 86)—the epigenetic projects of the early 2000s voiced a similar desire to make genomics responsive to the temporality of human life-courses, navigating between the inheritances of the past, and the exposure of individuals to the world and all its jostling complexities (Rando and Verstrepen 2007). Indeed, projects including O'Brien's (2007) examination of the intergenerational transference of PTSD symptoms, or Feil's (2006) analysis of environmental and nutritional effects on the epigenetic regulation of genes, spoke of expanded sphere of causation and temporality, where consequences lingered in uncanny ways, and influences ceased to be linear (Jablonka and Lamb 2006). Interest has instead concentrated in areas of enforcement and governmentality, where bodies are increasingly reduced to "their molecular components" (Rajan 2006, p. 163), leading to a reliance on "statistical, population-based data" (p. 163) and the intensified individualization of border processes over wider historic-contextual factors and communal stories (Benjamin 2009, p. 346).

This was salient in the Hildesheim case, where local authorities in the town of Hildesheim, Germany, were tasked with processing an asylum seeker who claimed to be under 18 years old, and thus, eligible for "exclusive" or privileged state services for unaccompanied minors (Abbott 2018). Due to the widespread narrative of asylum seekers exploiting such resources, and the accompanying air of suspicion, police turned toward a blood-based epigenetic age test provided by a privately held Californian company Zymo Research, which specializes in molecular biology research tools (Zymo Research Corp. 2018). The company had purchased an exclusive license to an epigenetic clock (DNAge™) developed by biostatistician Steve Horvath (from the University of California, Los Angeles), which measured methylation at 353 DNA sites across the genome (Horvath 2013).

The clock also offered the streamlined option to collect data via blood samples or via buccal cells scraped from the inside of the cheek. With Horvath using a pool of 53 people, aged between 3.5 and 18 years, he found a medium error of 1.03 years, with the figure taken enthusiastically up by Zymo as proof of the clock's evidentiary value and reliability when it came to genetically measuring maturity (Horvath and Raj 2018). Despite the quick circulation, claims of accuracy, and even the heralding of a legal utility (with Horvath publicly proclaiming that "the tests may help identify individuals who break the law" [Abbott 2018]), the paradata of outliers persistently haunt the epigenetic clock. Indeed, a study by Fleckhaus et al. (2017) revealed that Middle East population samples had significantly lower dispersions of methylations and were thus subject to higher rates of incorrect age certification than Central European and West African populations, while Horvath's (2013) own sample also had conspicuous deviations, with error rates up to 5 years and 8 months occurring.

While the technical unreliability of the tool itself presents an avenue of contestation, we have recourse to ask: what story do these devices communicate via their inclusion in border management projects like the one occurring in Hildesheim? Furthermore, what concrete bio-political realities are justified and materialized for those described through, and cast within, such narratives of *techno-biopower*? In his work Political Biology, Meloni describes the progression from hard heredity, at least in "the rigid version we have known it at the peak of genecentrism" (2016, p. 210), toward a "new softer synthesis," with the postgenomic soft genome supposedly possessing spatial and temporal dimensions "far more fluid and responsive to the environment than previously supposed" (Jablonka and Lamb 1995, p. 26). However, the attempts to manufacture predictable and trackable epigenetic clocks—from Horvath's to Hannum's clock (consisting of 71 markers that "accurately" estimate age based on blood methylation levels)—draw us back to age as a static, neutral object of inquiry, moving us along linear lifeways and trajectories despite the "analogical vastness" of environment variables (Meloni and Testa 2014). The conceptualization of the genome having "an early life and an old age, and to a more limited degree, an adolescence, middle age, and other stages" (Lappé and Landecker 2015, p. 152), while metaphorically compelling, still limits us to narratives of depoliticized and individualized biographies.

Horvath's clock acknowledges differential and unexpected paces of aging across life-courses—including instances of "age acceleration" (Horvath 2013), where epigenetic clocks exceed a person's chronological age. Nonetheless, they are generally restricted to medical reasoning, with such fluxes usually attributed to individually acquired and contained diseases—from neuropathology, Down Syndrome, Parkinson's disease, as well as select types of cancer (Perna et al. 2016). What this produces is an ontologically flattened personhood, where age ceases to be responsive to "different categories of things in the world" (Landecker and Panofsky 2013; Meloni and Testa 2014), ignoring the potentiality of the physical and physiological to be woven through with certain political realities—realities that attest to the effects of militarization and occupation, environmental racism, and the traumatization acquired in seeking asylum as having an epigenetic articulation (Meloni 2016). Such an alternate molecular narrative would force us to reckon with unchild-like children, or suspicious unaccompanied minors (Aitken 2001) who have inconsistent epigenetic clocks, as intimately entangled with contexts of violent maturation (McLaughlin 2018), brought on by interactions with exhausting forms of securitization and governance generally oriented toward hopelessness, stretched out states of anxiety, and the terrors of indeterminacy.

It is interesting to note that nonmigrant bodies receive radically different treatment in epigenetic research concerning aging. Within the West, there has been increasing interest in the discipline of biogerontology, as well as the emerging market of rejuvenation biotechnology (Mitteldorf 2016). Such areas hone in on the potential to medically treat aging via procedures such as senolytic therapy, where sénescent cells are removed under clinical conditions (Childs et al. 2017). Here, age is transformed into an approachable object—one could even say a *desirable* object which is open to improvement and subject to the innovation of biomedical science in reversing age-related decline (Schumacher 2018). In the case of nonmigrant bodies, therefore–bodies valued as subjects of a sovereign nation-state as well as potential consumers of anti-aging products–biomarkers become fabled indicators, which, if correlated with methylation processes, can be visibly targeted and transfigured (Benjamin 2009).

Horvath, in an interview with the Life Extension Advocacy Foundation and unpacking the metaphor of a clock (Milnova et al. 2018), muses on whether methylation is "part of the dial" or "part of the clockwork," invoking epigenetics as a salient tool for reworking temporality into more advantageous forms for those deserving, and/or those who can afford to invest in such technologies. Extending on this idea of temporal privileging and protection, it may be worth considering Rob Nixon's conceptualization of "enclaved temporalities" (Nixon 2011, p. 8). While his work specifically attends to the slow violences enacted in the context of environmental crises, and the ability for certain populations to live removed from "attritional lethality" (p. 8), there is a sense that epigenetic biotechnology also manifests a stratified zone of relations. For one group, it is mattering to be harnessed and manipulated for the sake of innovation as well as wellbeing; for another, it is simply matter to be extracted and used as traces of suspected criminality and otherness (Maguire and Rao 2018).

The chronopolitics of the migrant epigenetic clock is "somatechnically instrumentalized" (Pugliese and Stryker 2009, pp. 2–3), in the words of Pugliese and Stryker, reflecting a wider movement toward the bio-informationalization of select subjects and select lives in response to positivist science's demand "that nature reports itself in some way or other that is identifiable through calculation, and that it remains orderable as a system of information" (Heidegger 1977, p. 304). While these chrono-technologies promise fecund extensions and new contours of possibility for privileged subjects within the folds of techno-culture, migrant applicants, who encounter the very same interface (Randell-Moon and Tippet 2016), witness the rendering of their life into an orderable system of information. This system has the capacity to actively threaten the validity of anecdotal accounts; undermine existing documentation and produce versions of "embodied evidence" (Comaroff and Comaroff 2000, p. 298) which can accrue enviable value within the legal system, especially when wielded by the state to prove seeming inconsistencies and fraudulence (Maguire and Rao 2018). Indeed, there have been frequent instances within EU courts of documents–the personal testimonies of Afghan tazkiras, for example–being dismissed due to their "very simple quality," while the technics of dental and skeletal radiograms, anthropometry, and sexual development assessments were granted probative value and corroborated as supporting evidence (Hansson 2015).

This uptake of molecular evidence-making within sites of governance can be considered in relation to Eugene Thacker's outlining of the growing closeness between the biotech industry and a "larger context of globalization and political economy" (2005, p. xvii). It is this closeness which has oriented large-scale scientific endeavors—many of them concerned with the mapping of genetic information and the subsequent agglomeration of biological information into targeted databases—toward the production of globalized forms of biotechnology, where "economic exchanges, political exchanges, and semiotic exchanges are coupled with biological exchanges" (2005, p. xviii). It seems necessary to think of epigenetics in relation to contemporary political motivations, the desires of a bounded nation-state, and the intense scrutiny faced by bodies that move, cross, and challenge strict procedural forms as well as flows.

Estimated Nationalities, Calculated Politics in Border Science

The ascent of epigenetic age-verification technology dovetails with a more generalizable preoccupation around the unpredictability of physical, physiological, endocrinological, and psychological development (Aitken 2001), especially as it zig-zags across the arbitrary-set threshold between adult and child; in doing so exposing the complex embodiment of what it means to occupy the child-body (McLaughlin 2018). Alongside this, however, it is also a zone of conspicuous racialization, where race and ethnicity are both continuously reanimated through the effect of suspicion and the threat of a cheated system by those deemed outside of it (McGonigle and Benjamin 2016). Indeed, a common trope within mainstream media, when age-related cases leak out and attract attention, has been to circulate images of "unchildlike children"—children from predominantly Middle Eastern backgrounds, who may have more pronounced facial hair, be of heavier build, or taller than expected (McLaughlin 2018).

Articles with headlines such as "Adults are pretending to be children: Now even aid workers admit 'Calais kids' are LYING about their age" (Stevens 2016) and "Forever Young: Sweden Tussles With 'Refugee Children' in Their Thirties" (Sputnik 2016), as well as official findings from institutional bodies like the Swedish National Board of Medicines—which, allegedly, found that 442 cases out of the 581 assessments showed the particular minor in question to be 18 years or older (Sputnik 2017)—are entangled together to present maturity as firmly inscribed in molecular form, lying dormant within differently raced bodies, only to be exposed by proper scientific method (Lock 2015). It can be said that the potential use of epigenetics within border science promises to reify these caricatures by enacting digital and coded epidermalization (Browne 2009), whereby historic-racial schemas are reproduced on the level of the microscopic (Randell-Moon and Tippet 2016), stratifying race according to smaller and smaller units: from the dermal surface of the body down to single-nucleotide polymorphisms (SNPs).

With what Gilroy would describe as an intention "to make the mute body disclose the truth of its racial identities" (1997, p. 195), the recent trial of the Human Provenance Pilot Project (HPPP) speaks to the particular capacity of the epigenetic to animate new techno-material experiences of racial identity (Travis 2009). As an initiative by the United

Kingdom Border Agency (UKBA), the 2009 pilot project aimed to test the ancestry and geographical origins of asylum seekers applying for protection in the UK. The proposed method involved refugee subjects providing mouth swabs which would then be used for mitochondrial DNA and Y chromosome testing (Bohmer and Shuman 2018). Additionally, hair and nail samples were also to be utilized, in order to allow for the analysis of isotopes (Tutton et al. 2015). While isotopes of strontium and other heavier elements incorporate into bones and teeth throughout life, and thus provide some measure of geographic locales, the light element isotopes of hydrogen, oxygen, carbon, and nitrogen, which the Border Agency showed particular interest in, only hint at recent diets and climatic conditions (Balding et al. 2010). While it is interesting to think of a body as porous to the mundane matter which sediments and builds up in our local context, as well as the spaces of everyday life (Lock 2001), within the context of border control the emphasis tends to be on solidifying being, rather than acknowledging the intimate ways we "become" and are constantly "becoming" (Ingold 2013, p. 8).

Unlike the concept of embedded bodies envisioned by Margaret Lock, and that of local biologizes "designed to highlight inextricable multiplicities among material bodies and environments past and present, historical/socio/political variables, and subjectivities" (Lock 2015, p. 153), the epigenetics of nationality seeks to flatten race and ethnicity into official borderings which correspond to current territories of nation-states, and ideas of concrete nonwhiteness. As noted by David Balding, a population geneticist at Imperial College London, "genes don't respect national borders, as many legitimate citizens are migrants or direct descendants of migrants, and many national borders split ethnic groups" (Travis 2009), yet, despite this, as well as the reality of phenotypic difference being "highly dynamic and in constant flux" (Ramírez-Goicoechea 2013, p. 66), border agencies continue to hijack the lexicon of genomics in order to create new euphemisms for the continuation of intense racialization and exclusion. Indeed, it is coherent with a longer lineage of ontological insecurity experienced by nonwhite subjects, assuring that their bodies and bio-matter remain "out of place" (Browne 2009, p. 134). They are not afforded the destabilization of known parameters of identity and personhood—the type of flexibility and contestation environmental epigenetics was supposed to invite.

Instead, what migrant subjects experience is the very securitization of their genetic matter—a new repertoire where one's DNA can betray their very own autobiographical experiences and narratives (Lappé and Landecker 2015). Furthermore, with the increased reliance on biocentric jurisprudence, especially the re-hierarchization of material documents within the migration law as insufficient and untrustworthy when compared to scientific data (Luker 2016), we must ask what happens to already marginalized subjects in the legal system itself. Whereas state documentation, family photographs, and birth certificates allowed for an ability to forge counter-archives, enabled by the very materiality of documents, or "animated artifacts" in the words of legal scholar Trish Luker (2016, p. 71), the interpretation of the genetic by border agencies allows for no such subversions and resistances. The very extraction of the matter from the person, a taking away, also strips a potential agency to curate the material and imbue it with personal testimony (Maguire and Rao 2018).

Another example of note is the recent admission by Canadian immigration officials that their department was active in utilizing DNA testing and ancestry websites in order to try and establish the nationality of migrant applicants, particularly in cases where discrepancies had arisen (Khandaker 2018). This was encountered by Franklin Godwin, a Liberian man who had been the target of potential deportation after he acquired a criminal conviction, with the Canada Border Services Agency (CBSA) subjecting Godwin to a DNA test, as well as consulting various ancestry websites to see where his relatives were "actually from" (Hopkins 2018). In their prosecution, CBSA attempted to show that Franklin Godwin was originally from Nigerian, wielding a wrongful claim of nationality as proof of his delinquency and general deservingness to be a Canadian citizen. While the defendant's agreement is required for testing to occur, it should be noted that it is often under the threat of detention and noncompliance being used against them; a form of "extorted consent," as coined by lawyer Jared Will (Khandaker 2018).

These newly emerging procedures become representative of the desire to reinforce the boundaries of raced bodies, even on a molecular level, which threaten when they evade clear categorizations. It almost seems like the credibility of border policing (Bohmer and Shuman 2018) now relies on shutting down any avenues of self-description and self-representation, with the imaginary credibility of science trumping all other manifestations of evidence-making (Maguire and Rao 2018). It fits into a longer lineage of scientifically inscribing racialized bodies in order to speed up the efficiency of processing and to create systems of discrediting accounts. Indeed, as far back as migrant processing on Ellis Island during the 1900s, intense medical examinations circulated around the discovery of racial and ethnic identity, with such methods referred to as "sieves" that could effectively filter out unwanted and undesirable individuals (Fairchild 2003). They were, and continue to be, the "tricksters or fakes, assistants or toons" (Agamben 1993, p. 10), who, according to Agamben, are simultaneously the "exemplars of the coming community" (Agamben 1993, p. 10) threatening nation-states founded on fantasies of predictable genetic profiles and racial homogeneity.

Ultimately, as genetics is deployed to place more and more nonwhite and suspect bodies into detention and cycles of incarceration, bio-matter continues to circulate globally unaffected, with the biotech industry crosscutting the traditional boundaries of nation-states in its pursuit of capital and monetization (Thacker 2005, p. xvii). Certain nation-states will even strategically (re)biologize the cumulative genetic heredity of its population if a value is to be found via "biologically meaningful social groups" (Epstein 2007, p. 135). The uneven movement of the genetic material itself, and the radically differential types of personhood bestowed onto the molecular—from free-floating, cosmopolitan units of capital to demarcates of criminality and dishonesty—reveals that epigenetics can never exist in a depoliticized state, and its inclusion in border science begs for resistance and vocal opposition.

The Molecularization of the Asylum Seeker (The New Technics of Processing)

While the state seeks to create a "totalizing border environment" for "preemptive interventions on migrant movement through real-time data collection" (Vukov 2016, p. 84), we may come to recognize that the molecular personhood of migrants, when wielded by the individuals themselves, is imbued with an incredibly potent power—a power to shake and upturn the brittle stories which border control relies upon. Indeed, while the bodily, morphological, as well as molecular, has been time and time again configured and presented as credible evidence (Meloni 2016), the compelling alternative of the "body as reference, as enemy, as carrier of insecurities, as the agent of irritation, and the ever evading and liquid focus of regulation" (Maguire and Rao 2018, p. 19) exists, lingering as a constant threat to both border governance and securitization. Reflecting on the use of epigenetics for the verification of age and nationality in this chapter, what we may become aware of is that it simply doesn't tell the story of an exciting scientific development to come, which will salvage an increasingly fraught exercise in curtailing the movement of individuals across arbitrarily set territorial lines (Niewöhner 2011).

The bodies and molecular profiles of migrant's present gaps in knowledge and infrastructure, amplified by the inherent complexity of geographical origins and a distinct lack of reference populations against which epigenetic data can be cross-checked (Maguire and Rao 2018). Indeed, most medico-scientific methods—whether bone densitometry, carpal x-rays, or dental checks for an erupted third molar—employed by border agencies encounter very similar walls where the resounding issue of nontransferability haunts these technics. They simply cannot account for the dynamic zones of planetary life which individuals soak in, embody and absorb (Ahuja 2016).

Yet, these flattening *bio-polygraphs* persist—increasingly developed by transnational biotech firms; invested in by state departments; and enthusiastically taken up by border control bodies as well as a moralizing legal infrastructure where individuals are either deemed truthful or untruthful according to the data supplied. In the contemporary sphere ill-fitting ideas of genomic personhood are brought into existence within the space of the laboratory (Landecker 2011), where instances of social change are "operationalized" as "methylation, acetylation, and chromatic markers" simultaneously become seamless accounts of "abuse, poverty, stigma, violence, wars, and so on" (Lock 2015, p. 173). In this way populations become defined in ways that are "methodologically useful" for governance and administrative processing. As the laboratory stretches further and further into the sphere of the court (Lock 2015), molecules are imbued with certain representative agency, appearing as stable evidential facts within certain legal scripts. Indeed, in the U.S. this sort of intimacy between the attempt to prosecute and the justification for bioscientific methodology, is written into policy, with DNA collection and its subsequent entry into the national DNA database by law enforcement investigators only allowed if the suspect is affiliated with a potential crime (Kim and Katsanis 2013).

It is important to carefully watch for the subtle ways epigenetic knowledge, as well as a genomic lexicon, seeps into areas of border science and technology. How the desirable aspects of this new biomedical terrain also offer plentiful sites of exploitation and domination. It is a terrain rich with numerous designated clusters by which populations can be stratified: a terrain of genetically discernible groups that cover "ancestry, geography, 'folk' ethnoracial categories, and disease risk" (Benjamin 2009, p. 346), in its expansivity continuing a logic of concrete categorization and the solidification of otherwise complex and unruly subjects. It is also vital to be attentive to the ways people make sense of the epigenetic within our research in the humanities, as well as law, knowing that the particular techniques of storytelling employed can feed into predatory forms of knowledge production—especially as the value of the molecular is constantly mediated in the context of specific scientific, political, and economic institutions.

References

Abbott, A 2018, 'Can epigenetics help verify the age claims of refugees?', *Nature: International Journal of Science*, viewed 13 November 2018, https://www.nature.com/articles/d41586-018-06121-w.

Agamben, G 1993, *The coming community*, Minnesota University Press, Minneapolis, Minnesota.

Ahuja, N 2016, *Bioinsecurities: disease interventions, empire, and the government of species*, Duke University Press, Durham, NC.

Aitken, SC 2001, 'Global crises of childhood: rights, justice and the unchildlike child', *Area*, vol. 33, no. 2, pp. 119–127.

Balding, D, Weale, M, Richards, M & Thomas, M 2010, 'Genetic and isotopic analysis and the UK Border Agency', *UCL Molecular and Cultural Evolution Lab*, viewed 9 November 2018, https://www.ucl.ac.uk/mace-lab/publications/articles/2010/Balding_Signif10_UK_Border.pdf.

Benjamin, R 2009, 'A lab of their own: genomic sovereignty as postcolonial science policy', *Policy and Society*, vol. 28, no. 4, pp. 341–355.

Bohmer, C & Shuman, A 2018, *Political asylum deceptions: the culture of suspicion*, Palgrave Macmillan, Basingstoke, UK.

Browne, S 2009, 'Digital epidermalization: race, identity and biometrics', *Critical Sociology*, vol. 36, no. 1, pp. 131–150.

Childs, BG, Gluscevic, M, Baker, DJ, Laberge, RM, Marquess, D, Dananberg, J & van Deursen, JM 2017, 'Senescent cells: an emerging target for diseases of ageing', *Nature Reviews Drug Discovery*, vol. 16, no. 10, pp. 718–735.

Comaroff, J & Comaroff, JL 2000, 'Millennial capitalism: first thoughts on a second coming', *Public Culture*, vol. 12, no. 2, pp. 291–343.

Deleuze, G 1992, 'Postscript on the societies of control', *October*, vol. 59, pp. 3–7.

Epstein, S 2007, *Inclusion: the politics of difference in medical research*, University of Chicago Press, Chicago, IL.

Fairchild, AL 2003, *Science at the borders: immigrant medical inspection and the shaping of the modern industrial labor force*, Johns Hopkins University Press, Baltimore, MD.

Feil, R 2006, 'Environmental and nutritional effects on the epigenetic regulation of genes', *Mutation Research/Fundamental and Molecular Mechanisms of Mutagenesis*, vol. 600, no. 1, pp. 46–57.

Fleckhaus, J, Freire-Aradas, A, Rothschild, MA, Schneider, PM 2017, 'Impact of genetic ancestry on chronological age prediction using DNA methylation analysis', *Forensic Science International: Genetics Supplement Series*, vol. 6, pp. 399–400.

Genome Web 2018, 'HHS confirms DNA testing of migrant families with additional details', *Business & Policy*, viewed 14 November 2018, https://www.genomeweb.com/policy-legislation/hhs-confirms-dna-testing-migrant-families-additional-details.

Gilroy, P 1997, 'Scales and eyes: 'race' making difference', in S Golding (ed.), *Eight technologies of otherness*, Routledge, London, UK.

Glausiusz, J 2014, *Searching chromosomes for the legacy of trauma*, Nature Commentary, viewed 11 November 2018, http://www.nature.com/news/searching-%20chromosomes-for-the-legacy-of-trauma-1.15369.

Hansson, N 2015, 'Please state your age-An evaluation of the age assessment of unaccompanied minors seeking international protection in Sweden and whether such practice is in compliance with international and regional obligations', Masters thesis, Lund University, viewed 8 November, 2018, http://lup.lub.lu.se/student-papers/record/5434690.

Haraway, D 2016, *Staying with the trouble: making Kin in the Chthulucene*, Duke University Press, Durham, NC.

Heidegger, M 1977, *The question concerning technology and other essays*, Harper Torchbooks, New York, NY.

Ho, C 2018, *Effort to use DNA tests to reunite migrant families is paused*, San Francisco Chronicle, viewed 15 November 2018, https://www.sfchronicle.com/business/article/Effort-to-use-DNA-tests-to-reunite-migrant-13028878.php.

Home Office 2010, *FOI release: UKBA's human provenance pilot project*, GOV.UK, viewed 4 November 2018, https://www.gov.uk/government/publications/ukba-s-human-provenance-pilot-project.

Hopkins, A 2018, *Canada using DNA, ancestry websites to investigate migrants*, Reuters, viewed 6 November 2018, https://www.reuters.com/article/us-canada-immigration/canada-using-dna-ancestry-websites-to-investigate-migrants-idUSKBN1KH2KF.

Horvath, S 2013, 'DNA methylation age of human tissues and cell types', *Genome Biology*, vol. 14, no. 10, pp. 1–19.

Horvath, S & Raj, K 2018, 'DNA methylation-based biomarkers and the epigenetic clock theory of ageing', *Nature Reviews Genetics*, vol. 19, pp. 371–384.

IANS 2018, 'Canada border authorities use DNA tool to detect nationality', *The Week*, viewed 9 November 2018, https://www.theweek.in/news/world/2018/07/31/canada-border-authorities-use-dna-tool-to-detect-nationality.html.

Ingold, T 2013, 'Prospect', in T Ingold & G Pálsson (eds), *Biosocial becomings: integrating social and biological anthropology*, Cambridge University Press, Cambridge.

Jablonka, E & Lamb, MJ 1995, *Epigenetic inheritance and evolution*, Oxford University Press, Oxford, UK.

Jablonka, E & Lamb, MJ 2006, *Evolution in four dimensions: genetic, epigenetic, behavioral, and symbolic variation in the history of life*, MIT Press, Cambridge, MA.

Jones, R 2016, *Violent borders: refugees and the right to move*, Verso, London, UK.

Khandaker, T 2018, *Canada is using ancestry DNA websites to help it deport people*, Vice News, viewed 24 November 2018, https://news.vice.com/amp/en_ca/article/wjkxmy/canada-is-using-ancestry-dna-websites-to-help-it-deport-people.

Kim, J & Katsanis, S 2013, 'Brave New World of human-rights DNA collection', *Trends in Genetics*, vol. 29, no. 6, pp. 329–332.

Landecker, H 2011, 'Food as exposure: nutritional epigenetics and the new metabolism', *Biosocieties*, vol. 6, pp. 167–194.

Landecker, H & Panofsky, A 2013, 'From social structure to gene regulation, and back: a critical introduction to environmental epigenetics for sociology', *Annual Review of Sociology*, vol. 39, pp. 333–357.

Lappé, M & Landecker, H 2015, 'How the genome got a life span, new genetics and society', *New Genetics and Society*, vol. 34, no. 2, pp. 152–176.

La Spina, E 2017, 'DNA testing for family reunification in Europe: an exceptional resource?', *Migraciones Internacionales*, vol. 6, no. 22, pp. 40–75.

Lock, M 2001, 'Containing the elusive body', *The Hedgehog Review*, vol. 3, no. 2, pp. 65–78.

Lock, M 2015, 'Comprehending the body in the era of the epigenome', *Current Anthropology*, vol. 56, no. 2, pp. 151–177.

Luker, T 2016, 'Animating the archive: artefacts of law', in S Motha & H van Rijswijk (eds), *Law, memory, violence: uncovering the counter-archive*, Routledge, London, UK.

Maguire, M & Rao, U 2018, 'Introduction: bodies as evidence', in M Maguire, U Rao & N Zurawski (eds), *Bodies as evidence: security, knowledge, and power*, Duke University Press, Durham, NC.

Margulis, L & Sagan, D 2000, *What is life?* University of California Press, Berkeley, CA.

Mayr, E 1980, 'Prologue: some thoughts on the history of the evolutionary synthesis', in E Mayr & WB Provine (eds), *The evolutionary synthesis: perspectives on the unification of biology*, Harvard University Press, Cambridge, MA.

McGonigle, IV & Benjamin, R 2016, 'The molecularization of identity: science and subjectivity in the 21st century', *Genetics Research*, vol. 98, pp. 1–5.

McGowan, PO & Szyf, M 2010, 'The epigenetics of social adversity in early life: Implications for mental health outcomes', *Neurobiology of Disease*, vol. 39, no. 1, pp. 66–72.

McLaughlin, C 2018, 'They don't look like children: child asylum-seekers, the Dubs amendment and the politics of childhood', *Journal of Ethnic and Migration Studies*, vol. 44, no. 11, pp. 1757–1773.

Meloni, M 2016, *Political biology: science and social values in human heredity from eugenics to epigenetics*, Palgrave Macmillan, London, UK.

Meloni, M & Testa, G 2014, 'Scrutinizing the epigenetics revolution', *BioSocieties*, vol. 9, no. 4, pp. 431–456.

Mengiste, M 2018, 'This is what the journey does', in VT Nguyen (ed.), *The displaced: refugee writers on refugee lives*, Abrams, New York, NY.

Milnova, E, Hill, S & Conway, J 2018, *Steve Horvath—Aging and the epigenetic clocks*, viewed 21 November 2018, https://www.leafscience.org/steve-horvath-interview/.

Mitteldorf, J 2016, 'An epigenetic clock controls aging', *Biogerontology*, vol. 17, no. 1, pp. 257–265.

Molteni, M 2018, *Family DNA testing at the border would be an ethical Quagmire*, Wired Science, viewed 5 November 2018, https://www.wired.com/story/family-dna-testing-at-the-border-would-be-an-ethical-quagmire/.

Moore, D 2015, *The developing genome: an introduction to behavioural epigenetics*, Oxford University Press, Oxford, UK.

Nguyen, VT 2018, *The displaced: refugee writers on refugee lives*, Abrams, New York, NY.

Niewöhner, J 2011, 'Epigenetics: embedded bodies and the molecularization of biography and milieu', *BioSocieties*, vol. 6, no. 3, pp. 279–298.

Nixon, R 2011, *Slow violence and the environmentalism of the poor*, Harvard University Press, Cambridge, MA.

O'Brien, KJ 2007, 'The uncounted casualties of war: epigenetics and the intergenerational transference of PTSD symptoms among children and grandchildren of Vietnam veterans in Australia', Working Paper, Unpublished.

Perna, L, Zhang, Y, Mons, U, Holleczek, B, Saum, KU & Brenner, H 2016, 'Epigenetic age acceleration predicts cancer, cardiovascular, and all-cause mortality in a German case cohort', *Clinical Epigenetics*, vol. 8, no. 1, pp. 1–7.

Pugliese, J & Stryker, S 2009, 'The somatechnics of race and whiteness', *Social Semiotics*, vol. 19, no. 1, pp. 1–8.
Rajan, KS 2006, *Biocapital: the constitution of postgenomic life*, Duke University Press, Durham, NC.
Ramírez-Goicoechea, E 2013, 'Life-in-the-making: epigenesis, biocul- tural environments and human becomings', in T Ingold & G Pálsson (eds), *Biosocial becomings: integrating social and biological anthropology*, Cambridge University Press, Cambridge, UK.
Randell-Moon, H & Tippet, R 2016, *Security, race, biopower: essays on technology and corporeality*, Palgrave Macmillan, London, UK.
Rando, O & Verstrepen, K 2007, 'Timescales of genetic and epigenetic inheritance', *Cell*, vol. 128, pp. 655–668.
Salter, F 2017, *On genetic interests: family, ethnicity and humanity in an age of mass migration*, Routledge, London, UK.
Schumacher, A 2018, 'Epigenetics of aging and longevity: challenges and future directions', in A Moskalev & AM Vaiserman (eds), *Epigenetics of aging and longevity*, Academic Press, Cambridge, MA.
Sputnik 2016, *Forever young: Sweden tussles with 'refugee children' in their thirties*, Sputnik News, viewed 11 November 2018, https://sputniknews.com/europe/201609051044983695-sweden-age-verification-refugee-children/.
Sputnik 2017, *It's official: most refugee 'children' in Sweden revealed to be adults*, Sputnik News, viewed 17 November 2018, https://sputniknews.com/europe/201705311054149872-sweden-refugee-children-adults/.
Stevens, J 2016, 'Adults are pretending to be children', *Daily Mail*, viewed 16 November 2018, https://www.dailymail.co.uk/news/article-3847114/Adults-pretending-children-aid-workers-admit-Calais-kids-LYING-age-vulnerable-nine-year-old-refused-UK-entry-shambolic-selection-process.html.
Thacker, E 2005, *The global genome: biotechnology, politics, and culture*, MIT Press, Cambridge, MA.
Travis, J 2009, 'Scientists Decry "flawed" and "horrifying" nationality tests', *Science*, viewed 19 November 2018, http://science.sciencemag.org/content/326/5949/30?casa_token=73mKBT2vgbAAAAAA:O8wCFanDKg8wfl4IsGk8jYsBHz4KHTZKn5VMRf4o2AH7iebjubbxQnMxNKvf4sh7PyZK9J_VSGec0w.
Tutton, R, Hauskeller, C & Sturdy, S 2015, 'Importing forensic technologies into border control: genetic ancestry and isotope testing in the UKBA's Human Provenance Pilot Project', in BN Lawrance & G Ruffer (eds), *Adjudicating refugee and asylum status: the role of witness, expertise, and testimony*, Cambridge University Press, New York, NY.
Vukov, T 2016, 'Target practice: the algorithmics and biopolitics of race in emerging smart border practices and technologies', *Transfers*, vol. 6, no. 1, pp. 80–97.

Zhang, S 2018, 'Which company will test the DNA of separated families?' *The Atlantic*, viewed 7 November 2018, https://www.theatlantic.com/science/archive/2018/07/when-dna-tests-are-used-to-reunite-immigrant-families/564584/.

Zymo Research Corp. 2018, *Zymo research's epigenetic aging clock service (DNAge™ Test) to be featured on Quarks & Co. TV Show*, CTSTUN PR Newswire, viewed 22 November 2018, https://www.prnewswire.com/news-releases/zymo-researchs-epigenetic-aging-clock-service-dnage-test-to-be-featured-on-quarks–co-tv-show-300580072.html.

CHAPTER 9

CRISPR Cowboys? Genetic Self-Experimentation and the Limits of the Person

Courtney Addison

Introduction: Hacking the Future

Let's begin with two visions of the future. The first comes courtesy of Elizabeth Parrish (2017b), who writes:

> Those humans alive 200 years from now will have exquisite control over their health and wellbeing through advanced cell technologies that can only be imagined by people living today. The risk takers - those who can already envision surpassing the limits of their bodies and those who are unwilling to accept the disability of old age without a fight - will be the ones pioneering a bold future and reinventing our concept of health.

It appears that a lot is about to change. The second vision comes from Josiah Zayner, who conjures up an alternative in an interview with the *Guardian*:

C. Addison (✉)
Centre for Science in Society, Victoria University of Wellington
Wellington, New Zealand
e-mail: courtney.addison@vuw.ac.nz

M. de Leeuw and S. van Wichelen (eds.),
Personhood in the Age of Biolegality, Biolegalities,
https://doi.org/10.1007/978-3-030-27848-9_9

> To me it's like Bladerunner, where he goes into that backalley science lab and there's the guy making eyes. I imagine people going to some place like a tattoo parlour, and instead of getting a tattoo they pick out some DNA that makes them muscly, or changes the colour of their hair or eyes. DNA defines what a species is, and I imagine it wouldn't be too long into the future when the human species almost becomes a new species because of these modifications. (Ireland 2017)

A bold future, a new species, our concept of health reinvented. This rhetoric is familiar in the context of the life sciences, of what we used to call the new genetics, and certainly, still, of the genomic era. It is a genre of hype that we are accustomed to hearing from biotech executives, market analysts, and science journalists. And indeed, Parrish and Zayner are both executives: Parrish is CEO of Bioviva, and Zayner CEO of the Odin. But these two differ from their more conventional counterparts, because both have put their bodies on the line. Parrish and Zayner are part of a niche group of genetic self-experimenters.

Genetic self-experimenting describes the act/s of administering self-developed gene constructs to one's own body. Some of the people who engage in these practices call themselves biohackers, but "conventional" biohackers (if that isn't too great an oxymoron) are often offended by the stunt-like execution of some of this activity. Most biohackers focus on dietary and behavioral "hacks" (for example nootropics, sleep cycle manipulations) to improve or extend their bodily capacities. A smaller sector, "grinders," physically modify their bodies using, for example, magnets and biometric chips. In contrast, the people I'm calling genetic self-experimenters are using DNA, and they are using it primarily for one of two purposes: to try to cure diseases, or to overcome aging.[1] Parrish is in the latter group. Her company, Bioviva, aims to develop a gene therapy for aging, and Parrish was patient zero in the company's trial (there does not appear to be a patient one or two).

This chapter focuses on the contact zone between self-experimentation and anti-aging. Here, a novel image of human bodily possibility emerges, and a new proposition for the social landscape of science as well. At stake in genetic self-experiments are individuals' own futures, which they want to extend, and the future role of science, which they seek to transform. I analyzes genetic self-experiments' public accounts, drawing on journalistic material and "first person" media such as YouTube videos and Reddit commentary. This material centers on a handful of individuals, all of whom are located in the United States.[2] Its geographic location and cultural referents make genetic self-experimentation typical of Silicon Valley science, another permutation of the "California Ideology" described by Barbrook and Cameron (1996). I am particularly interested in the confluence of genetic and digital technologies, which coalesce around genetic self-experimentation in a way that gives (certain) bodies the power to travel and signify anew, even as the fleshier transformations sought of them prove hard to realize. The limits of the person, in this context, are thus both bodily and social. Accounting for persons here means accounting for the power relations that produce or undo subject positions, and the multiple ways in which persons and bodies can affect and be affected through new technological mediations.[3]

A Brief History of Self-Experimentation

Contemporary organized medical research and its institutions (for example Institutional Review Boards [IRBs], regulators) arose after World War II, partly in response to atrocious Nazi science (Annas 2009). Prior to this, research had been the domain of individuals, often working outside formal institutions. Early experimentation "was generally observational and uncontrolled" (Kerridge 2003, p. 203). A review spanning 1800–2000 notes a peak in self-experimentation in the first half of the twentieth century (Weisse 2012), primarily for infectious diseases, but also in anesthesiology, physiology, and pharmacology (ibid.). During the latter nineteenth and earlier twentieth centuries, the medical profession was largely self-regulating. It was widely agreed that their professional standards and expertise made doctors themselves the best overseers of their practice, and trust in the institution of medicine ran deep (Wilson 2014).

The experiments medics and scientists have carried out on themselves range from haphazard to systematic (Kerridge 2003). Self-experimentation led to a number of prominent medical advances, and seven Noble prizes have been awarded for findings derived from this work. Weisse (2012) notes that these experiments were usually undertaken either to establish a baseline of "normal values" against which to compare future findings, or to demonstrate a given procedure's safety to colleagues. It is plausible, however, that there was also an affective impulse behind this activity. Weisse (2012, p. 54) describes "the feeling of camaraderie that one experiences in contributing to a research enterprise and in partaking, even a little, in the adventure of discovery." Sympathetically, Kerridge (2003, p. 206) diagnoses "an insatiable curiosity and a need to participate closely in [one's] own research."

Accounts of self-experimentation tend to reproduce the (male) scientist-as-hero genre.[4] Brilliance and bravery are often foregrounded, and these narratives tend to individualize the knowledge production process. Virologist Tom Rivers (Benison in Weisse 2012, p. 54) claimed that:

> great discoveries are not made by committees or groups of workers; they originate in the minds of single individuals... I know of no important discovery in medicine or biology in the last 100 years that evolved out of gang research. You can do a hell of a lot of scut work by gang research, but the ideas for discovery are still going to come from the ideas in one man's mind

Work from the history of science and feminist STS has, of course, shown that "discovery" is rarely an individual process, and masculinist narratives of scientist-as-hero reinforce an unhelpfully gendered image of science (c.f. Haraway 1994; Terral 1998). Weisse (2012) notes that Jonas Salk (developer of the polio vaccine) and A.B. Sabin (who later developed an oral version), both tested their vaccines on their wives and children as well as themselves. The aggrandizing lens of "self"-experimentation mispresents the multiple bodies and persons from whom those findings derived.

Self-experimentation waned around the middle of the 1900s, as new codes of conduct and external oversight were brought to bear on the profession. However, we see something of a renaissance in the recent rise of biohacking and genetic self-experimentation. This new flourish of activity is enabled in part by the ready availability of cheap molecular biology tools, and the relative ease with which lab techniques can be learnt online.

Contesting the Human: Contemporary Self-Experimentation

Self-experimentation has long occupied an ambiguous position in relation to ethics, law, and regulation. United States regulation does not account for self-administered investigational products. Indeed, United

States law permits a wide range of behaviors that are detrimental to one's own health or well-being (for example smoking, self-harm). Where those behaviors impinge on the health or safety of others—that is, where individual liberty comes into conflict with the social good—laws or regulations may be enacted to safeguard societal interests by curtailing individuals' rights (c.f. Oriola 2009, for an example of this in relation to smoking).

The US Food and Drug Administration (FDA) oversees the sale and regulation of medical products in the United States and is responsible for determining and then enforcing that these meet acceptable standards of safety and efficacy. However, the agency's remit does not extend to overseeing the use of those products by an individual acting alone. Regulating their sale and movement of these products thus becomes one way for authorities to exert a measure of control over the likes of self-experimentation. For example, in a statement titled "Information about the self-administration of gene therapy," the FDA (2017) sought to limit the sale of Zayner's product:

> FDA is aware that gene therapy products intended for self administration and "do it yourself" kits to produce gene therapies for self administration are being made available to the public. The sale of these products is against the law. FDA is concerned about the safety risks involved. Consumers are cautioned to make sure that any gene therapy they are considering has either been approved by FDA or is being studied under appropriate regulatory oversight.

This statement stresses the illegality of selling products for self-experimentation. However, it also captures the FDA's inability to prohibit the personal use, by an individual, of modified genetic technologies, and we see the language of the statement slip from legality to caution.

Self-experimentation ethics are similarly murky. The Nuremberg Code holds that "no experiment should be conducted when there is an a priori reason to believe that death or disabling injury will occur; except, perhaps in those experiments where experimental physicians also serve as subjects." This would appear to exempt self-experimenters from oversight. The 1974 National Research Act mandated the creation of IRB tasked with protecting human research subjects' rights. What qualifies as human subject's research, however, appears undecided. Ethics boards differ in their estimations of whether individual, self-administered research warrants review. In one example, an IRB declined to review a proposal as it did not meet their definition of human subject's research, but the same project later required approval from a different institution's IRB when they sought to share their data with that institution's researchers (Landhuis 2016). In a case report in which an author is denied publication of his $n = 1$, self-administered study, The Committee on Publication Ethics (COPE) advise that he has the right to carry out the research, but no corresponding right to be published (COPE 2015).

Commercial factors have also allowed genetic self-experimentation to flourish. It has become increasingly easy to purchase and learn to use genetic tools online. This is partially due to advances in gene editing technologies: Zinc Fingers and Clustered Regularly Interspaced Short Palindromic Repeats (CRISPR) can now be obtained at relatively low-cost. Vendors such as Zayner sell these in the form of "kits" that give consumers the tools and information they need to use them at home (although Zayner's company, the Odin, stopped selling such kits for human use following the FDA statement above, they now sell and ship a version containing live frogs). The commercial and regulatory landscape of the United States is such that individuals can readily buy the necessary equipment for genetic experiments and can use them on themselves without contravening the law.

Performing Self-Experimentation

Elizabeth Parrish, whose quotation opened this chapter, is the charismatic CEO of Bioviva, and by her own word, "the first patient to be treated with gene therapy to reverse aging" (Parrish 2015). Parrish is engaging and professional. Although she has no scientific training, she surrounded herself with people who do. Her company is advised by people who have Ph.D.s and work at Universities or research institutes. Bioviva has all the trappings of a standard biotech firm, with a Scientific Advisory Board, a "technology platform," and a fancy website. However, they do not entirely follow the conventions of traditional medical research: Parrish's trial took place in a clinic in Colombia, deliberately out of reach of the FDA, and neither Bioviva nor the administering medical team had performed the preclinical groundwork for their first-in-human test, working instead off published results of research conducted elsewhere. The event was reportedly captured on film for a documentary, and this film has been viewed by at least one journalist (see Funk 2018). It has also been narrativized by Parrish herself, who hosted an Ask Me

Anything "AMA"[5] on Reddit about the experiment and has had numerous speaking engagements and media interviews about it since. Her portrayal of the procedure is at once dramatic ("I am patient zero. I will be 45 in January. I have ageing as a disease" [Parrish 2015]) and pragmatic ("We are using both visual biomarkers, Magnetic resonance imaging [MRI] and a panel of blood tissue testing including work on telomere length with Spectracell and Life Length epigenetic testing" [ibid.])

Aaron Traywick pursued rather different means to a similar end. Recently deceased (for reasons not directly related to his experiments),[6] Traywick gained international attention in early 2018 when he was streamed on stage at a biohacking conference, whipping off his pants and injecting his thigh with what he called "the world's first gene therapy-based vaccine and cure for herpes". Traywick, like Parrish, had no scientific training (Brown 2018), but got together with other DIY biologists to form Ascendance Biomedical. According to its website, the company aims to find cures for HIV and herpes as well as aging. Traywick, though, was not the first recipient of an Ascendance Biomedical gene product.

In 2017, a young man called Tristan Roberts was live streamed sitting on a couch, in between Traywick and biohacker Machiavelli Davis. Roberts has HIV and had become disillusioned with the anti-retroviral he was taking to keep it in check. When he discovered Ascendance online, he enrolled as the first patient to receive their "gene therapy"[7] for HIV. Via a Facebook live stream, 160 people watched Roberts inject the contents of a vial containing the N6 gene into himself. The procedure did not work, though it didn't cause Roberts any physical harm either. He has since resumed his medication, and Ascendance Biomedical has collapsed. The initial live stream of Roberts's procedure produces a particular kind of biohacker aesthetic, as viewers see this group of young men handling scientific instruments in a space that is equal parts grungy and domestic. Sometime later Vice made a documentary about the group. Serendipitously shot as Ascendance fell apart, the film takes viewers through the group's community lab space and shows them arguing in a low-lit bar.

Another self-experimenter, Brian Hanley, is CEO and founder of the one-man enterprise Butterfly Sciences. Hanley developed a construct based on the Growth hormone-releasing hormone (GHRH) gene, which codes for a growth regulating hormone. An application to a private IRB stated his aim as to increase the hormone to "more youthful levels" (Regalado 2017). Hanley, who is also a biologist, enrolled a plastic surgeon to administer the construct, and recorded the event on video. Unlike most gene therapists, who use viruses to shuttle their gene construct into the body, Hanley used a naked DNA plasmid, which was effectively shaken into his cells through a process called electroporation: an electrical current was passed through his body, creating a ripple effect that momentarily loosens up the cell membrane, allowing the passage of the plasmid. A scientist who was asked for comment said that when he was told about Hanley's method, "I thought they had to be kidding. It's usually something we do to animals" (Regalado 2017). As in Parrish's case, the administration event was filmed, though there are no reports of it having been viewed or shared. Hanley was profiled in the MIT Technology Review following his procedure (ibid.).

The performativity of these events is striking. Whether in the form of a live stream or the real time back and forth of an Ask Me Anything (AMA), the internet enables genetic self-experimenters to speak and perform directly to an audience. The high drama of these events is amplified by the cultural power of the gene (Nelkin and Lindee 1995), with its associations of potency, risk, and fundamental humanity. At the same time, the nature of the internet is such that a person in one country can sit in their living room and experience the happenings–however unusual–of another person's living room elsewhere in the world.

An Experiment to End All Aging?

For Parrish, a future without age is a natural but radical extension of previous medical innovations: she observes that antibiotics and vaccinations have granted a quality of life that was unimaginable 200 years ago. Espousing a very linear vision of scientific progress, Parrish advocates for this momentum, for us to continue pushing out the horizon of what is biologically possible.

This relies on a particular view of aging, as primarily biological, and as morally wrong. (Telomere biologist Bill Andrews: "A lot of people claim that aging is something that we can do gracefully. It's not. Aging is one of the worst things that can happen to a person" [Funk 2018].) In a talk, Parrish (2017a) recounts her gene therapy procedure, and asks:

> Why would I take two gene therapies when medicine today would consider me healthy? I did it because I believe I am dying of a disease. And so are you. That disease is biological aging.

"Biological ageing" is easily read as a medicalization of what was previously a complex biosocial process, and here as elsewhere medicalization functions to create the kind of biology that needs fixing (Conrad 1992). This unfolds in a broader context of preventative care, intensive monitoring, and the invention of "biological age" (Moriera 2015). The fact that companies like Bioviva turned first to gene therapy to "treat" aging suggests a molecularizing optic as well. As Bioviva's website explains:

> …as cells age they become increasingly poor at functioning and repairing damage and this is why tissues and organs fail and we get disease… We are developing gene therapies that target cellular aging and encourage cellular and tissue regeneration. Among the various causes of aging such as cellular senescence, accumulated waste and changes to gene expression, are some of the leading risk factors of aging and disease.

Here, aging is genetic and cellular. The body is under-realized, awaiting the kind of interventions that can extend, improve, and enhance its existing capacities.

Their fixation on aging as disease positions self-experimenters between the biohacking community and more "mainstream" anti-agers. Anti-aging science has become an enormous market, with a flurry of activity in America's prime innovation hotbed, Silicon Valley. Tech tycoons including Amazon's Jeff Bezos, Facebook's Mark Zuckerberg and Priscilla Chan, and Apple's Arthur Levinson are investing in or founding significant anti-aging ventures. For a purported $8000/liter, Bay area start-up Ambrosia will sell you literal young blood, bought from people younger and hopefully healthier than yourself.

Harvard molecular biologist George Church is quoted in the Washington Post imagining a future in which, "everyone takes gene therapy" … for "diseases that everyone has, like ageing" (Achenbach 2015). Tapping a well-entrenched narrative, Church calls the "ageing population" one of "our biggest economic disasters" (Achenbach 2015). Rather than seeking a social or economic solution, though, anti-agers favor developing medical products that can prolong our life spans. Attempts to profit off age-related worries are nothing new (c.f. Petersen 2018), but molecularizing age brings it into the remit of genetic medicine, where the financial stakes and bodily risks are stark. This view of aging advances a neoliberal idea of the person as self-improving project, and the body as a critical substrate of that improvement. It also points to how medicalizing and capitalizing logics encroach ever further into the body.

Radical Rhetoric

The performativity of much genetic self-experimentation is paired with a similarly performative rhetoric. Resting on critiques of research and governance institutions, this rhetoric extolls values of openness and transparency, which are pitched as antithetical to current research systems. Genetic self-experimenters employ grandiose and oppositional language, portraying themselves as radical agents of change–or, in Silicon Valley parlance, disruption.

This group takes issue with institutions from the FDA to universities and ethics boards. They often frame these as slow-moving and obstructive, if not dangerous. Before sticking the syringe into his belly, Tristan Roberts looks at the camera and dedicates the act to "all the people who have died while not being able to access treatment even though those treatments were available" (Fischer 2017). Of course, frustrations with access to medicines are neither new, nor unjustified. For this group, however, they come to justify entirely different models of medicines access.

A magazine article portrays Parrish recounting the hurdles of conventional medical research:

> …"to bring gene therapies to you," Parrish went on, "I would have to go raise almost a billion dollars. It would take about 15 years of testing. And when I'm looking out there, I'm seeing people who don't want to wait 15 years." The crowd began clapping, and Parrish fed off it. "How do we actually change this paradigm? Well, what we do is we burn and raze everything to the ground. And we start over." (Funk 2018)

Of course, it is not just the costs of developing a medical product that is impeding Bioviva's research. It is also the fact that while the medical establishment recognizes a wide range of age-*related* diseases (as set out, for example, in the International Classification of Diseases [2018]) as appropriate targets of medical intervention, they do not regard aging in itself as a disease. Her comments, though, resemble Traywick's. Not long before his death, Traywick explained, "There are breakthroughs in the world that we can actually bring to market in a way that wouldn't require us to butt up against the FDA's walls, but instead walk around them" (Bowen 2018). The journalist here, Kristen Brown (2018), notes that this evasion is enabled by what their website calls:

> "a pioneering decentralised research paradigm." That paradigm, as it turns out, relies mostly on volunteers to perform experiments on themselves... Biohackers, some of whom have no formal scientific education at all, help the company do the lab work for cheap. According to the company 'concept paper', cryptocurrency is somehow also involved.

These parties also frame bioethics as unnecessarily limiting. Parrish (2017b), for example, argues:

> Bioethics are steeped in hesitation. Today's bioethics have become overly cautious by only considering the potential harm that could be caused by an experimental treatment, and not taking into account the number of people dying while waiting for new treatment.

This portrayal neglects questions of beneficence that are central to bioethical reasoning. However, it is unsurprising that the field attracts this kind of commentary. Bioethics has long been integral to biomedical research (Jasanoff 2007), and it is the terms of biomedical research that self-experimenters take issue with. In fact, the entire of rationale of bioethics would appear to be not so much obstructive to self-experimenters, as irrelevant. The field is predicated upon the fact that some people (researchers, doctors, medical professionals) have control over the pain, life, and death of others (patients, research subjects). Here, the research process is individuated, collapsing the researchers-subject relationship. If we view the person as single, bounded and continuous, then when the practitioner and subject are the same person, their duty of care is arguably void.

Hanley exemplifies this dilemma. He gained approval from an independent IRB but did not disclose that he was his own research subject. "I am informed consent personified," he argued. "There is no one in the world more informed than me" (Regalado 2017). Hanley undoes the usual social format of medical research, problematizing the subject role. If Hanley the research subject answers to Hanley the researcher, then the usual power relations of scientific research become redundant. As I will argue, however, this reshuffling of roles does not truly individualize the research process, nor does it eliminate the problematic power relations of medical research.

In contrast to the norms of contemporary medical and research ethics, self-experimenters extoll openness and transparency, ideals they share with DIY biology and open science movements (Delfanti 2013; Wexler 2017). These values animate the who's and how's of self-experimentation. Church's Harvard lab often runs tests and provides advice to independent self-experimenters, including Hanley and Ascendance biomedical. When asked if he felt that providing this service encouraged risky experiments, Church demurred:

> We certainly don't encourage people to do this; in fact, we encourage them not to […] I don't think of it as offering a haven so much as a critique. (Regalado 2017)

Josiah Zayner offers a similar perspective:

> …if these tools aren't made available, people will still make them somehow. Pushing it to the outskirts, pushing it underground, is going to push people to do ridiculous uninformed things. If it's out in the open, I can have people coming to me and asking questions about it. You have people coming to the community asking questions. (Zhang 2018)

Church and Zayner agree that openness permits those with more expertise (or self-restraint) to act as conscience or counsel to those who might try high-risk activities.

There is a libertarian streak in these accounts. "Why should anyone block me from doing with my body what I wish to do with it?" asks Parrish (2017b), "aesthetics, tattoos, clothing choices or gene therapy-it is my body, my choice." Elsewhere, she is reported urging an audience, "You have a right to do with your body what you wish!" (Funk 2018). This implies a false dichotomy between individuals' rights to their bodies, and the role of biomedical expertise in mediating between those bodies and experimental medical products. Her comments resonate with the Ascendance groups. In Roberts' self-experiment video, Traywick announces, "It is his [Roberts'] body, and it is his right and his freedom to be able to do research as he sees fit." The third group member introduces himself: "My name's Machiavelli Davis. I support human freedom. I'm a biohacker."

Discussion: The Limits of the Person

As the Ascendance group set up on-camera, Traywick stated that their product was not for human consumption, and that they did not advise anyone to copy the procedure they were about to broadcast. "Tristan

Roberts," Traywick states, "is completely within his rights according to the FDA and the rule of law in his nation and ah in most other nations around the world… to self-experiment on his self in any way that he deems medically appropriate. It is his body, and it is his right…." Roberts was clearly an enthusiastic volunteer, and the process itself exploited the regulatory lacuna around self-administered experimental products. However, that lacuna implies an individualization and autonomy that this scene troubles. The social makeup of these self-experiments is not reducible to whose hand holds the syringe, though this is the strategy Ascendance adopts to ensure they are within the law. To focus only on this detail misrepresents the many dependencies and relationships that make these "self"-experiments possible. Roberts (syringe in hand) is connected to a live audience of unknown viewers, to the Ascendance team seated around him (who have financial stakes in "his" experiment), and to more distant companies and suppliers who sold the group equipment and tested their therapy.[8]

The regulatory gap that allows self-experimentation can be critiqued on the same grounds as the concept of the autonomous individual in research ethics. The autonomous individual is the human keystone of biomedical research ethics, which require a subject who can make decisions and freely offer consent. But critics argue that the individual itself is a cultural artifact (Das 1999), and autonomy a shaky concept at best (Conrad 1992). Autonomy is continually compromised by people's social commitments, bodily frailties, and the power dynamics of medicine. The same is true of "self"-experimentation.

These events also problematize the idea of bodies as bounded, that one can self-administer an experimental medicine without affecting others. That idea works off a mechanistic, biologized view of the body, devoid of semiotics or meaning. Roberts may not shed genetically modified material that will affect other people or organisms (a concern in early gene therapy, and to some extent still today). But the procedure was live streamed and is searchable with a few keywords punched into Google. Traywick's caveats suggest he too knew that Roberts' procedure could affect the bodies of untold viewers; it belies his awareness that Ascendance is a party to that influence. Ironically, the immortality of the internet makes bodies like Roberts' all the more "persuasive" (Gardner et al. 2019), even as the bodily perpetuity that these actors strive for proves elusive. The physical body may be moored in space and time, but its image travels and endures in ways that empower it to speak beyond its limits.

In some cases, these self-experimenter's bodies also travel materially. Parrish's procedure was performed in Colombia, and before his death Traywick was reportedly planning a trial in Mexico. While Parrish helped devise and realize the gene therapy, she did not administer it; the process, involving multiple injections over several hours, was performed by local medical practitioners. This highlights another important aspect of this small group: they represent a privileged few, who have the resources to relocate themselves in relation to national regulatory environments. In many ways this is a mere extension of existing patterns, which see clinical trials offshored *en masse* to the Global South. There remains a marked distinction between who stands to profit from radical new science, and whose resources are relied upon to do so; what counts as radical here? Furthermore, despite their anti-establishment rhetoric there seems to be less of an appetite for eliminating the authoritative posts of existing gatekeepers, and more a desire to unseat the existing "core set" and replace them with a new guard. Instead of the FDA and IRBs, we now have Church and Zayner, sober advisors to enthusiastic amateurs.

The limits of the person are being recast by a technological confluence of genetic and digital tools. These limits are at once biological (how long can we live?), social (who can we reach?), and regulatory (what can we do to ourselves and to others?). Such questions pose a challenge for the law, and how it should address the power relations of self-experimentation. Hanley, Parrish, and peers seek to circumvent existing research relations by positioning themselves as both subject and researcher. In such an arrangement there is no expert mediation between at-risk recipients of a gene construct and its designer. How can regulators cope with this new terrain, where traditional arbiters of ethical practice and scientific expertise have been pushed out of the picture? More fundamentally, how can the need for regulatory oversight be reconciled with changing subject roles and the enduring right to exercise agency over one's body? If the nature of personhood is brought into flux with developments in the biosciences, we need to stay mindful of the fact that it is not only persons as recognized social beings that are at stake, but also subjects–persons positioned in relation to power. How this relationship between persons, subjects, and power is articulated remains to be seen.

Notes

1. Others are self-experimenting with genetic tools for other purposes. Justin Atkin, for example, self-developed and administered a gene therapy to improve his lactose intolerance.
2. See Delfanti's (2013) account of the United States' particular history of lay biology.
3. Thanks to Hallam Stevens, who provided generous and thoughtful feedback on the talk that this chapter developed from.
4. For insight into what an anti-heroic turn might look like, see Kasprzack (2016).
5. 'AMA' stands for Ask Me Anything, and is a convention of the web forum Reddit. Public figures will host an AMA in which Reddit users can put questions to them, and receive real time responses, for a set period of time.
6. Traywick drowned in a floatation tank with recreational Ketamine in his system.
7. The scare quotes around gene therapy are intended to indicate that this is Ascendance's language. Critics of the gene therapy field proper have pointed out that the language of therapy misrepresents what is in fact an experimental procedure that often does not have any therapeutic effects whatsoever (Churchill et al. 1998).
8. Anna Wexler (2017) has astutely pointed out that even the most fringe biohackers are still reliant on the mainstream scientific communities, whose suppliers, methods, and data they use to inform and support their own experimentation.

Bibliography

Achenbach, J 2015, 'A Harvard professor says he can cure ageing, but is that a good idea?' *The Washington Post—Achenblog*, web blog post, viewed 18 May 2018, https://www.washingtonpost.com/news/achenblog/wp/2015/12/02/professor-george-church-says-he-can-reverse-the-aging-process/?noredirect=on&utm_term=.89b3b31a211a.

Annas, G 2009, 'The legacy of the Nuremberg doctors' trial to American bioethics and human rights', *Minnesota Journal of Law, Science and Technology*, vol. 10, no. 1, pp. 19–40.

Barbrook, R & Cameron, A 1996, 'The California ideology', *Science as Culture*, vol. 6, no. 1, pp. 44–72.

Becker, G & Nachtigall, RD 1992, 'Eager for medicalisation: the social production of infertility as a disease', *Sociology of Health & Illness*, vol. 14, no. 4, pp. 456–471.

Bowen, KV 2018, 'What does an infamous biohacker's death mean for the future of DIY science?', *The Atlantic*, viewed 10 May 2018, https://www.theatlantic.com/science/archive/2018/05/aaron-traywick-death-ascendance-biomedical/559745/.

Brown, K 2018, 'This rogue company wants people to inject themselves with untested drugs', *Gizmodo Australia*, https://www.gizmodo.com.au/2018/02/this-rogue-company-wants-people-to-inject-themselves-with-untested-drugs/.

Churchill, R, Collins, L, King, NMP, Pemberton, SG & Wailoo, KA 1998, 'Genetic research as therapy: implications of "gene therapy" for informed consent', *Journal of Law, Medicine and Ethics*, vol. 26, pp. 38–47.

COPE (Committee on Publication Ethics) 2015, *The ethics of self-experimentation (Case Number 15-03)*, Retrieved from https://publicationethics.org/case/ethics-self-experimentation.

Conrad, P 1992, 'Medicalization and social control', *Annual Reviews of Sociology*, vol. 18, pp. 209–232.

Das, V 1999, 'Public good, ethics, and everyday life: beyond the boundaries of bioethics', *Daedalus*, vol. 128, no. 4, pp. 99–133.

Delfanti, A 2013, *Biohackers: the politics of open science*, Pluto Press, London, UK.

FDA 2017, 'Information about self-administration of gene therapy', U.S. Food and Drug Administration website, viewed 20 May 2018, https://www.fda.gov/BiologicsBloodVaccines/CellularGeneTherapyProducts/ucm586343.htm.

Fischer, F 2017, 'Biohacker self-administers attempt at gene-therapy HIV cure', *News2Share*, Facebook update, https://www.facebook.com/N2Sreports/videos/biohacker-self-administers-attempt-at-gene-therapy-hiv-cure/1557343577706859/.

Funk, M 2018, 'Liz Parrish wants to live forever', *Outside Online*, https://www.outsideonline.com/2325556/liz-parrish-live-forever.

Gardner, J, Warren, N, Addison, C & Samuel, G (2019), 'Persuasive bodies: testimonies of deep brain stimulation and Parkinson's on YouTube', *Social Science and Medicine*, vol. 222, pp. 44–51.

Gieryn, T 1983, 'Boundary-work and the demarcation of science from non-science: strains and interests in the professional ideologies of scientists', *American Sociological Review*, vol. 48, no. 6, pp. 781–795.

Haraway, D 1994, 'A game of cat's cradle: science studies, feminist theory, cultural studies', *Configurations*, vol. 2, no. 1, pp. 59–71.

Ireland, T 2017, 'I want to help humans genetically modify themselves', *The Guardian*, viewed 10 May 2018, https://www.theguardian.com/science/2017/dec/24/josiah-zayner-diy-gene-editing-therapy-crispr-interview.

Jasanoff, S 2007, *Designs on nature: science and democracy in Europe and the United States*, Princeton University Press, Princeton, NJ.

Kasprzack, M 2016, 'From innovator to maintainer: the anti-heroic turn', *EASST Review*, vol. 35, no. 4.

Kerridge, I 2003, 'Altruism or reckless curiosity? a brief history of self-experimentation in medicine', *Internal Medicine Journal*, vol. 33, pp. 203–207.

Landhuis, E 2016, 'Do it yourself? when the researcher becomes the subject', *Science News*, viewed 5 December 2016, https://www.sciencemag.org/careers/2016/12/do-it-yourself-when-researcher-becomes-subject.

Moriera, T 2015, 'Unsettling standards: the biological age controversy', *The Sociological Quarterly*, vol. 56, no. 1, pp. 18–39.

National Research Service Award Act 1974. 'Public Law 93–348', Retrieved from https://history.nih.gov/research/downloads/PL93-348.pdf.

Nelkin, D & Lindee, M 1995, *The DNA mystique: the gene as a cultural icon*, University of Michigan Press, Ann Arbor.

Oriola, T 2009, 'Ethical and legal analyses of policy prohibiting tobacco smoking in enclosed public spaces', *Journal of Law, Medicine & Ethics*, vol. 37, no. 4, pp. 828.

Parrish, E 2015, '[AMA] my name is Liz Parrish, CEO of BioViva, the first patient to be treated with gene therapy to reverse aging, ask me anything', *Reddit*, 11 October 2015, https://www.reddit.com/r/Futurology/comments/3ocsbi/ama_my_name_is_liz_parrish_ceo_of_bioviva_the/.

Parrish, E 2017a, 'Liz Parrish: ageing is a disease: gene therapy could be the cure' (video), *Wired Events*, viewed 10 May 2018, https://www.wired.co.uk/video/ageing-disease-gene-therapy-cure.

Parrish, E 2017b, 'One year anniversary of Bioviva's gene therapy against human aging', viewed 12 May 2018, https://bioviva-science.com/blog/2017/3/2/one-year-anniversary-of-biovivas-gene-therapy-against-human-aging.

Petersen, A 2018, 'Capitalising on ageing anxieties: promissory discourse and the creation of an "anti-ageing treatment" market', *Journal of Sociology*, vol. 54, no. 2, pp. 191–202.

Regalado, A 2017, 'One man's quest to hack his own genes', *MIT Technology Review*, viewed 10 May 2018, https://www.technologyreview.com/s/603217/one-mans-quest-to-hack-his-own-genes/.

Terrall, M 1998, 'Heroic narratives of quest and discovery', *Configurations*, vol. 6, no. 2, pp. 223–242.

Weisse, A 2012, 'Self-experimentation and its role in medical research', *Texas Heart Institute Journal*, vol. 39, no. 1, pp. 51–54.

Wexler, A 2017, 'The social context of "do-it-yourself" brain stimulation: neurohackers, biohackers, and lifehackers', *Frontiers in Human Neuroscience*, vol. 11, pp. 224.

Wilson, D 2014, *The making of British bioethics*, Manchester University Press, Manchester.

World Health Organisation. (2018). *International Classification of Diseases, 11th Revision (ICD-11)*. Retrieved from https://icd.who.int/browse11/l-m/en.

Zhang, S 2018, 'A Biohacker regrets publicly injecting himself with CRISPR', *The Atlantic*, viewed 10 May 2018, https://www.theatlantic.com/science/archive/2018/02/biohacking-stunts-crispr/553511/.

CHAPTER 10

In Genes We Trust: Genetic Privacy in the Age of Precision Medicine

Dean Southwood

Introduction

We live firmly in the Age of Big Data. The creation and popularity of large-scale data-generating tools has inundated many areas of science and society with data, both personal and impersonal. This trend, combined with the current flourishing Age of Biotechnology, has created a flashpoint of research, crafting the field of genomics into a scientific powerhouse. Genetic sequencing techniques are rapidly improving, and on-demand whole genome sequencing (WGS) at the individual level is on the horizon. A shift toward precision medicine (also known as personalized medicine)—that is, medicine tailored on an individual level to each patient, based on their genetic condition—is already happening, and is likely to become increasingly relevant as current technology continues to improve. The private sector has already made the leap to direct-to-consumer genetic testing (DTC-GT), in an area which is significantly less regulated, and subject to far less institutional caution than healthcare. A move toward electronic health records (EHRs) in many

D. Southwood (✉)
University of New South Wales, Sydney, NSW, Australia
e-mail: dean.southwood@unsw.edu.au

M. de Leeuw and S. van Wichelen (eds.),
Personhood in the Age of Biolegality, Biolegalities,
https://doi.org/10.1007/978-3-030-27848-9_10

jurisdictions also forces us to reconsider our relationship with data, its protection, and exactly how much privacy we are willing to give up.

The rapid growth of these technological areas has, either directly or indirectly, created big expectations. Precision medicine in particular promises significant improvements in public health at the individual level and is marketed by many as an essential technological advancement (Salari and Larijani 2017). The meteoric rise of individual WGS data, both from the healthcare sector as well as the private sector, is likely to have significant ramifications, from research into genetic diseases and biomarker discovery (Kichko et al. 2016), to use by forensic geneticists as evidence in criminal trials (Scudder et al. 2018). However, maintaining positive public perception and trust in such technology hinges on correct management of the fundamental issues which arise (Chow-White et al. 2015). In particular, issues around the right to know (or not to know) certain genetic information have come to the fore, dragging with it a conversation around scientific literacy, and exactly how much information should be given with genetic test results. Should spurious, statistically insignificant findings still be presented to patients in the name of full disclosure, or is this ethically irresponsible? In addition, issues around access, storage, security, and use of genetic data are still causes for concern. Whether employers and insurers have access to the results of genetic tests is an ongoing discussion in many jurisdictions, and standards for correct handling and secure storage of data are still in development. The level of consent required from patients is also wildly different across jurisdictions and brings with its discussions around how positive consent must be, for what uses of data, whether consent can be revoked, and potential proprietary rights maintained over the data.

To properly address these issues, there needs to be a collaborative, productive dialogue between the legal and the scientific communities, to determine what the desired outcomes of precision medicine, EHRs, and DTC-GT schemes are, what bounds need to be placed on these technologies, and what management schemes and privacy policies are required on the data produced. From a science and technology studies standpoint, the interplay between law, society, and science in this area is an apt illustration of the idea of the coproduction of law and science as put forward by Jasanoff (2007, p. 772). Due to the complexity of the technology involved, and the far-reaching effects of its use, collaborative approaches that are firmly interdisciplinary are the most likely to produce satisfactory outcomes which take into account the concerns of primary stakeholders. Legislation and restrictions need to be placed before the generation and use of data becomes too widespread to regulate, highlighted strongly in the social media sphere, where companies such as Facebook have become almost too big to regulate effectively. Efforts in the genetic privacy sphere need to be swift but deft in order to manage a field promising many benefits, but also risking many problems.

This chapter is divided into six sections. "Legal Personhood and Ethical Questions" begins with a discussion of fundamental ethical questions surrounding precision medicine and big data, including a discussion of how our concept of legal personhood is expanding in this new Age. "Precision Medicine, Electronic Health Records, and Genetic Testing" explores the current state of precision medicine and genetic testing, including a brief overview of the technologies available, as well as desirable outcomes that are likely to be upcoming milestones in the future. "Genetic Privacy and Precision Medicine: Current Framework" surveys current frameworks governing precision medicine and genetic privacy more broadly, across a number of common and civil law jurisdictions including Australia, the European Union, and the United States. "Genetic Privacy and Precision Medicine: Open Questions" discusses key open questions that still plague the rise of precision medicine, centered around genetic privacy and genetic testing, highlighted through two key questions: is there a right to know (or not know) our genetic information; and who else has access to such information, and for what purposes? These questions are explored in "The Right (Not) to Know" and "Access, Use, Consent, and Security" respectively. "Conclusion" then provides future directions of inquiry and highlights what areas of reform must be focused on collaboratively by science and law in the near future.

Legal Personhood and Ethical Questions

The concept of legal personhood is in many ways one of utility. The law must interact with particular legal actors in definite ways, whether it be through the conferring of rights, the placement of duties, the granting of legal standing, or the judgment of criminality. In order to codify interactions, the law deems certain entities "sufficient" to fall under its purview, while other entities are found lacking. While the starting point of such ideas may have been that human beings (and only human beings) are persons, the concept has evolved to encompass many non-human actors; the legal personhood granted to corporations is the typical example used to abstract the concept away from a definable individual. However, in the new Ages of Biotechnology and Big Data, the concept has again been stretched to attempt to cover new ballooning rights and responsibilities, and in doing so has once again abstracted even further.

Through our interactions with ever increasing data collection mechanisms, the volume of our personal digital data is becoming substantial. Data marketing companies are already using digital "crumbs" of data to reconstruct digital versions of a person (Ebeling 2016, p. 29). However, these are inherently imperfect—the data used is not always accurate, is often provided for other purposes than it is being used for and is not comprehensive across many relevant issues. Therefore, what is created is almost an alter ego of the person, one which is used by many other legal actors in ways that almost attribute to it a legal personality. Several questions arise from this: what legal rights and responsibilities are associated with this new entity? Can they be considered a legal person? What does this then mean for how society and law are structured? These questions are particularly relevant in the face of precision medicine, where fundamental pieces of personal data are collected for this exact purpose: to create a digital medical double, which can be interrogated to diagnose and treat the "original" person more effectively. These ideas feed into other conceptual areas, such as how property rights can be defined in this area. There has been debate around whether we think of people as "manifestations of genetic data that are appropriately mined and propertized, or as entities that are outside the realm of commercial use" (Bonython and Arnold 2015, p. 378). As data miners and data brokers rise in prominence, the possible practical options in this scenario become less and less, with the former approach seeming to succeed in the absence of, but also in the presence of, detailed governance (Ebeling 2016, p. 75). These are fundamental questions which need to be probed and answered collaboratively, with input and cooperation from actors in the scientific sphere, the commercial sphere, the legal sphere, and the political sphere, in order to affect a viable outcome.

Precision Medicine, Electronic Health Records, and Genetic Testing

The upcoming field of precision medicine, also known as personalized medicine, has grown in relevance in recent years. The rapid advancement and expansion of genetic sequencing and other genomic tools has put the assembly of individual genomes squarely on the horizon. Cheaper, more portable technologies are also in rapid development, meaning the sequencing of a patient (a) at their bedside, (b) with sufficient accuracy, and (c) for under $1000 is fast approaching reality. This section deals with the current state of such sequencing technology, as well as the scientific landscape surrounding the use of such sequencing data—what does precision medicine promise to accomplish? This section also examines two related areas to precision medicine—EHRs, which are being rolled out in a number of jurisdictions, and DTC-GT, as offered by companies such as Ancestry.com and 23andMe among others. The data generated from such services are already readily available and serve as a guide for how precision medicine will be dealt with. While precision medicine is still a number of years away from being broadly available, EHRs and DTC-GT are already here and are therefore key in determining what concerns must be addressed.

Precision medicine, also known as personalized medicine (a term which has somewhat fallen out of vogue with public policy agencies) or individualized medicine in some jurisdictions, promises to improve medical outcomes by tailoring treatment strategies based on individual data, primarily genetic sequencing data. The field has been on the horizon for a number of years but has risen in the public eye recently due to a number of factors, both social and scientific. There have been recent public announcements of funding of precision medicine research, in particular the Precision Medicine Initiative in the U.S. announced by President Obama (White House 2015) and funded by U.S. Congress in 2016. As part of this initiative, the All of Us Research Program aims to take data from at least 1 million voluntary participants over the course of 10 years (National Institutes of Health 2019). In addition to public initiatives, scientific advancement has also fueled speculation on the state of precision medicine. Genetic sequencing technology has progressed substantially in recent years, with a key (somewhat arbitrary) metric being passed—it is now possible to sequence an individual genome for under $1000 (Lightbody et al. 2018). Sequencing technology is becoming increasingly accurate, with higher coverage of complex regions of the human genome possible with "long-read" technology such as those offered by PacBio SMRT Sequencing and Oxford Nanopore Technologies ("ONT") sequencing. Sequencers are also becoming more portable—ONT sequencers such as MinION and PromethION are pocket-sized, with the vision of bedside individual genome sequencing a very near reality.

Precision medicine is usually defined in a broad manner, encompassing a myriad of different techniques and outcomes, usually centered around genetic sequencing data. The most common technologies included under the umbrella are the identification of biomarkers of disease and cancer, the tailoring of drugs to individual genetic makeups for increased potency, and the design of treatment programs which suit a particular genetic condition rather than a general group. The use of genetic data for paternity testing and genealogy (often referred to as genetic genealogy) is highly related, although less commonly referred to as precision medicine. However, these secondary data uses are in fact the most common in today's society, and therefore serve as a case study for how data processing and handling will likely be managed as precision medicine advances.

Other technologies such as EHRs and DTC-GT are also acting as significant disruptors. EHRs are electronic records of healthcare data, allowing for sharing and syncing of data across multiple healthcare appointments, institutes, and practices, and are being rolled out in a number of different jurisdictions to mixed effect (Wang et al. 2017; Mendelson 2018). As repositories of personal health data, including increasing amounts of genetic data, they are a hot source for researchers interested in large-scale individual genomes, as well as other interested third parties such as law enforcement and employers. Schemes to facilitate research access to deidentified health data are being run in the U.S., such as the Electronic Medical Records & Genomics Network (eMERGE), in order to allow for scientific development, while also mitigating privacy risks. The scheme is opt-in, and only a small number of institutions have opted to collect data so far, but such schemes are likely to continue to grow in the future, warranting further investigation into how they function. In the private sector, DTC-GT companies such as Ancestry.com and 23andMe have been operating for a number of years, offering consumers a sequenced version of their genome. Depending on the company, different services are provided, including likelihoods for certain disease markers, ancestry information, as well as raw sequences. As a highly unregulated market in terms of privacy and data management, particularly outside of the European Union, which in essence provides personal health information to consumers, the ramifications of such an industry also require further study to identify and mitigate issues.

While these technologies promise improvements in health outcomes, higher efficiency, and novel services, there are a number of issues that still need to be mitigated, both legally and ethically. In a legal context, issues around privacy and data protection are still significant, with a lack of comprehensive legislation in most jurisdictions. Genetic data is highly personal, so much so that many researchers argue that it requires special treatment. Such "genetic exceptionalism" relies on a number of key distinctions about genetic data; firstly, it has the possibility of creating significant ramifications for insurance and employment if made public; secondly, it is unchangeable, and is accessible long into the future with the same ramifications; thirdly, the potential uses for such data in the future are highly uncertain, with many uses likely yet to be thought of, let alone developed; and fourthly, genetic data can impact beyond simply the individual being sequenced—it can provide information about (genetic) family members as well (Miller and Tucker 2017, p. 4651). There are also lingering issues around informed consent, particularly when genetic data is used for research purposes, in which case many future applications of the data are unknown; it is therefore difficult to properly consent to such use in advance. Questions of who can access genetic data outside of the healthcare sector are also potent—a study of population groups in the U.S. found that the majority of individuals are significantly concerned about insurers, law enforcement, and employers being able to access the results of genetic testing (Clayton et al. 2018). The law surrounding such access is still developing, but genetic information is already available through DTC-GT services, often being able to be distributed in unregulated ways, unknown to the consumer (Hazel and Slobogin 2018).

Ethical questions are also still unanswered, including what information should be provided to recipients of genetic information—what right do patients have to know spurious and possibly insignificant results? In a similar way, what right do patients have to not know such genetic information? Do patients have a right to refuse to disclose such information to family members? What about medical practitioners, are they bound by confidentiality to not reveal such genetic information to family members who may be at risk? There are also more fundamental questions around the idea of precision medicine and its role in the social frameworks in which we live. Due to its likely high cost, is this yet another stratification of healthcare on a socioeconomic scale, where one can "buy" better health in even more ways? (Prainsack 2017) Even more fundamentally, precision medicine in some ways presents a technological solution which masks a social problem (Prainsack 2017, p. 8). For marginalized communities, mortality rates, and access to healthcare are substantially different in many countries, particularly the U.S. (Canedo et al. 2018). These questions emphasize the continuing dialogue that needs to take place alongside medical developments to retain the focus on key social and ethical issues.

The current framework surrounding pressing legal questions is dealt with in section "Genetic Privacy and Precision Medicine: Current Framework" below, analyzed across a number of jurisdictions; the ethical and open legal questions are then discussed in section "Genetic Privacy and Precision Medicine: Open Questions".

Genetic Privacy and Precision Medicine: Current Framework

The need to craft legislation to deal with genetic privacy has been evident for over a decade. Due to the advances in genetic sequencing technology in the last decades of the twentieth century, as well as the sequencing of the human genome at the turn of the twenty-first century, the idea that a person's genetic information may be available for scrutiny has been present in both the public and legal zeitgeist, across multiple jurisdictions. However, the issue has been dealt with in different ways by different jurisdictions, with no jurisdiction completely capturing the breadth and gravity of the issues on the horizon. It is valuable however to examine what has already been done and what the current state of play is, in order to illustrate what must still be done to mitigate such problems before they arise. Therefore, this section will deal with significant relevant legislation and law reform across Australia, the European Union, and the United States, which concerns genetic privacy and genetic sequencing data.

The Australian Law Reform Commission (ALRC) tabled a report in 2003, commissioned by the Howard Government, which deals with primarily with issues of genetic privacy (Australian Law Reform Commission 2003). The report sets out 144 recommendations, including the modification of guidelines for ethics committees, the establishment of an advisory body to oversee protections and progress in human genetics in Australia, through to changes to the Privacy Act 1988 (Cth) (No. 119, 1988) (Privacy Act); very few of these recommendations were implemented in any form. The Privacy Act definitions of "health information" and "sensitive information" have been amended in 2006 to include the reference to genetic information (Privacy Act, s. 6); however, there is no definition of "genetic information" provided, leaving the interpretation of such a key provision an open question. The Privacy Act also includes a provision which allows for disclosure and use of genetic information under certain conditions (s. 16B [3]), which may include for research purposes (Eckstein et al. 2018, p. 586). Most guidelines for genetic testing results, and any consideration of data protection, consent, and ethics, are all contained in an advisory document provided by the Australian Government National Health and Medical Research Council (2018) entitled the National Statement for Ethical Conduct in Human Research (National Statement). While this has been updated in recent years to take into account technological advancement, it is still lacking in comprehensive advice on data storage and protection measures, as well as disclosure obligations, creating uncertainty for many researchers working in the area (Eckstein et al. 2018, p. 584). Australia also has little regulation in terms of insurer and employer use of genetic data—in fact, many insurers are already using this data. However, at present, there appears to be little change in premiums, and the lack of regulation is providing useful data about how insurers go about using genetic data when it is available for access and use (Prince 2018). However, this is a particular quirk of the Australian jurisdiction, one which likely requires legislative intervention in the near future before the availability of genetic information becomes too big to effectively regulate, in order to prevent genetic discrimination.

Data and privacy regulation in the European Union have been significantly reformed in recent years, primarily through the passing of the General Data Protection Regulation (EU 2016/679) (GDPR). This has radically changed the landscape of data storage and use in the European Union, with many areas of the regulation still yet to be litigated on, and therefore interpretationally uncertain (Mostert et al. 2018). In particular, there has been a shift in emphasis from "privacy" to "data protection"—the resulting legal effect of such a shift is as-yet unknown. Genetic data is dealt with specifically by the GDPR, as well as more broadly applicable issues such as informed consent, what can be done with the data once it is collected, and how it must be stored and transferred appropriately. Particularly relevant to the research community is the definition of "pseudonymized" data, which has been deidentified, but the ability to recreate such links is still within reasonable means (Shabani and Borry

2018, p. 151). For much research work on human samples, this is the case—this data is still deemed to be "personal data" and is therefore subject to strict requirements. Informed consent provisions are often perilous for genetic research on human samples—Article 8 of the GDPR requires informed consent for disclosure and use unless it is of substantial public interest; Article 8 Recital 34 provides that scientific research is of substantial public interest, conditional on "suitable safeguards" (GDPR EU 2016/679, art 8).

Areas of precision medicine and genetic testing in the U.S. are currently covered at the federal level by the Health Insurance Portability and Accountability Act (HIPAA) 1996, as well as the Genetic Information Nondiscrimination Act (GINA) 2008. By operation of the two, employers and insurers are prohibited from using genetic information as a basis for determining premiums or making employment decisions. However, the legislative future of such provisions is uncertain (Wang et al. 2017). The research use of EHRs is regulated through the Health Information Technology for Economic and Clinical Health Act (HITECH) 2009, allowing for the emergence of the Electronic Medical Records and Genomics Network (eMERGE). This allows for research institutes to opt-into a data sharing program which is regulated, ideally maintaining privacy while allowing for open research progress. The U.S. situation is also somewhat complicated: under HIPAA, individuals have a civil right to obtain all health information from healthcare providers, including full genomic testing results. This raises issues of interpretation, statistical significance and scientific literacy (Evans 2018); however, there are efforts underway to try and alleviate such problems by creating online tools to help the general public interpret genetic sequencing data (Curnin et al. 2017). Further guidelines on what information must be provided to individuals alongside their genetic sequencing data, including ways to "make sense" of the data, as well as guidance to doctors who may receive patients concerned about their genetic data, would likely help alleviate much of the potential issues.

Due to an increasingly globalized system of corporations and data sharing, it is anticipated that jurisdictions will clash on issues such as genetic privacy and data protection. In the Australian case, such a clash is arising from issues of consent and data use for EHRs (Mendelson 2018). The "My Health Record" system is the Australian version of EHRs, seeking to create individual electronic versions of patient health records in order to facilitate information transfer and consistency across the health system, as well as to prevent patients from "double-dipping" on prescriptions. However, a number of privacy issues have arisen surrounding the scheme, which has yet to be fully rolled out, and is currently an "opt-out" system.

Many jurisdictions have some framework concerning genetic information and genetic privacy; however, there are still significant gaps in legislation, regulation, guidelines, and interpretation, leaving many important issues still up-in-the-air. Reform is required particularly in the Australian context to remedy the situation, mitigate against looming problems, and attempt to future-proof as best as possible against the abuse of genetic information.

Genetic Privacy and Precision Medicine: Open Questions

The current frameworks concerning genetic privacy and precision medicine are at worst incomplete, and at best unclear and without court interpretation. Therefore, there are still pressing open questions surrounding the data collected from genetic testing, from precision medicine, direct-to-consumer services, and electronic health information available on EHRs. This section will explore the questions surrounding two broad issues: the right to know (or not to know) your genetic information; and access, use, consent and security of genetic data. There are many other issues which must be dealt with as soon as possible: broader concerns surrounding genetic discrimination, data sharing across jurisdictional lines, as well as larger discussions around research guidelines are key examples (Majumder 2018). These still require in-depth discussion in order to minimize their impact on existing social and ethical tensions and to come to substantially better solutions than are currently presented; however, this section will hone in on two key issues, in an attempt to form a framework for discussion going forward.

The Right (Not) to Know

There is some conceptual confusion around the idea of "the right to know" one's genetic information, and alternatively "the right not to know". Laurie (2014) identifies that these are not merely "two sides of the same coin"—there are different interests at stake in both cases. Therefore, both will be considered separately.

In the case of the right to know, this relies on a basis of autonomy, and once genetic sequencing has been done, there is an obligation to disclose such information to the individual being sequenced. However, the pertinent question then becomes: disclose what? The nature of the genetic information provided to the test recipient is debatable and speaks to a deeper divide between science and law as truth-makers. Science provides caveats and conditions on its truths, never speaking in absolutes—interpretation of science as it is done, particularly in the life sciences, requires a significant level of statistical literacy, which cannot be expected of the public at large. However, in legal terms, truth is decided absolutely—it is, or it is not. When the two fields clash, this dissonance must be resolved. Therefore, in the case of genetic sequencing results, the regulation requires a standard for what is disclosed, particularly in the U.S. where there is a civil right to full disclosure of genetic information from testing (Evans 2018). Some guidelines have been developed on what genetic markers to disclose in reports, and what interpretation information should be disclosed (Frizzo-Barker et al. 2016). However, there is a real possibility for some patients of having "hypochondrial genetic panic", in misunderstanding the risks associated with a certain genetic makeup. Being a highly technical discipline, it is therefore the responsibility of the researchers, healthcare providers, and sequencing operators to provide adequate information about the results obtained, especially in the case of research where results may arise which were not asked for—for example, when looking for markers for autism, the researchers find that the individuals may also be prone to a genetically-based cancer—should the patients be notified, if the evidence is not 100 percent certain (Eckstein and Otlowski 2018, p. 403).

On the other hand, the right not to know one's genetic information is harder to enforce. It is difficult to define a way in which such a right could be held against an agency seeking to give a person their genetic information—as argued by Laurie (2014, p. 40), the act of asking if they wish to know is giving them some information already. Cowley (2016) argues that in the modern age of genetic disclosure and responsibility to genetic family members, as well as a societal shift toward "genetic responsibility" in general, the right not to know is slowly being eroded. This is a ground for ethical discussion upon which reasonable people are likely to differ—guidelines should be in place to deal with the ethical dilemmas which may arise from a person's refusal to know their genetic information, such as when a doctor treats both a genetic test recipient, and their immediate genetic relative—if the recipient refuses to disclose evidence of a genetic illness to their relative themselves, can the doctor breach confidentiality to disclose this to the relative themselves? Current guidance suggests they should if the risk is significantly high, but this illustrates the ripeness of the area for ethical dilemmas (Suthers et al. 2011).

Access, Use, Consent, and Security

The main legal and societal questions surrounding genetic testing and EHRs concern access to genetic data-who can access it, what consent is required, what can it be used for, and what security measures must be in place to protect the data? In terms of access, there are a number of parties who would wish to obtain genetic information, namely law enforcement, employers, insurers, and researchers. In the case of law enforcement, forensics would ideally like to be able to cross-reference a DNA sample found at a crime scene against a DNA database of possible suspects; in addition, techniques in forensic genomics are also advancing, painting a picture of a suspect through their genetics alone (Scudder et al. 2018). In recent weeks, such forensic genomics has been used in the U.S. to identify a suspect 26 years later, due in part to a relative submitting their genetic information to a public genealogy database (Scolforo 2019). However, there are issues around using information that the suspect may not know—would disclosing that in the course of a criminal investigation be ethical? In jurisdictions such as The Netherlands where such techniques have been more routinely employed, there have been strict guidelines on what markers can and cannot be used to predict suspects.

In the case of employers and insurers, there are issues of genetic discrimination—if an individual is highly likely to develop a genetic disease, should they be charged a higher insurance premium? If they have a genetic predisposition that makes them unable to complete certain work, can an employer fire them on such grounds? This has been dealt with in a number of jurisdictions, through anti-discrimination legislation or voluntary moratoriums by insurance companies; however, other jurisdictions such as Australia leave open these possibilities (Prince 2018).

Issues surrounding consent and data protection are still very much open questions. The implementation of the GDPR in the European Union has sparked debate over whether legislation should be passed in other jurisdictions, further placing controls on how data is managed, what consent must be obtained for its use, and how it must be protected. As much of the GDPR has yet to be litigated, time will tell how effective it proves to be; in the meantime, guidelines in other jurisdictions such as Australia must be put in place to prevent consent abuses, such as those likely to arise from the "My Health Record" scheme (Mendelson 2018).

Conclusion

The rapid advancement of genetic sequencing technology has created a suite of possibilities across many fields of science, as well as across society more broadly. However, in an age where personal data are becoming increasingly more valuable and increasingly more prone to exploitation, the regulation surrounding the privacy, access, and use of such data must be robust and able to mitigate major issues before they arise. In almost all jurisdictions, there is insufficient legislation, regulation, guidelines, and legal guidance on what must be done in the healthcare sector, the research sector, and the private sector, in order to maintain the privacy and security of personal data. In Australia in particular, there is significant work left to be done. While regular amendments to the National Statement are useful in guiding researchers, more comprehensive reform is necessary. The *Essentially Yours* report (Australian Law Reform Commission 2003) requires updating to include technological advancements, and a more comprehensive handling of the private sector as well as the Internet. The major recommendations of the report still need to be implemented, in particular surrounding the use of genetic information by insurance companies, which is currently all but handled on the basis of goodwill. As data sharing continues to increase, and multinational companies continue to handle genetic data, international consensus will be required on how best to regulate these activities. Stringent data protection regulation in the EU in the form of the GDPR, which has a reach into European companies' activities internationally, will require other jurisdictions to decide how best to fit their own data protection schemes around it. Further detailed conversations need to be had, both between the legal and scientific community as well as between experts and the public, in order to highlight key issues and find ethical, satisfactory solutions to them. To ensure the further success of medical technology advancement, public engagement is essential. Such conversations also need to happen as soon as possible, in order to preempt major issues before they become crises, or before practices become too ingrained to easily curb. By way of a collaborative coproduction of law and science, the major issues presented by precision medicine, EHRs, and DTC-GT can be mitigated, and the benefits of such technological advancements enjoyed at a significantly lower risk to our personal freedoms.

References

Australian Government National Health and Medical Research Council 2018, *National statement on ethical conduct in human research*, viewed 1 April 2019, https://www.nhmrc.gov.au/about-us/publications/national-statement-ethical-conduct-human-research-2007-updated-2018.

Australian Law Reform Commission 2003, *Essentially yours: the protection of human genetic information in Australia*, no. 96, viewed 1 April 2019, https://www.alrc.gov.au/publications/report-96.

Bonython, W & Arnold, B 2015, 'Privacy, personhood, and property in the age of genomics', *Laws*, vol. 4, no. 3, pp. 377–412.

Canedo, J, Miller, S, Schlundt, D, Fadden, M & Sanderson, M 2018, 'Racial/Ethnic disparities in diabetes quality of care: the role of healthcare access and socioeconomic status', *Journal of Racial and Ethnic Health Disparities*, vol. 5, no. 1, pp. 7–14.

Chow-White, P, MacAulay, M, Charters, A & Chow, P 2015, 'From the bench to the bedside in the big data age: ethics and practices of consent and privacy for clinical genomics and personalized medicine', *Ethics and Information Technology*, vol. 17, no. 3, pp. 189–200.

Clayton, EW, Halverson, CM, Sathe, NA, & Malin, BA 2018, 'A systematic literature review of individuals' perspectives on privacy and genetic information in the United States', *PLoS One*, vol. 13, no. 10, e0204417.

Cowley, L 2016, 'What can we learn from patients' ethical thinking about the right 'not to know' in genomics? Lessons from cancer genetic testing for genetic counselling', *Bioethics*, vol. 30, no. 8, pp. 628–635.

Curnin, C, Gordon, A & Erlich, Y 2017, 'DNA Compass: a secure, client-side site for navigating personal genetic information', *Bioinformatics*, vol. 33, no. 14, pp. 2191–2193.

Ebeling, M 2016, *Healthcare and big data*, Palgrave Macmillan, Philadelphia.

Eckstein, L, Chalmers, D, Critchley, C, Jeanneret, R, McWhirter, R, Nielsen, J, Otlowski, M & Nicol, D 2018, 'Australia: regulating genomic data sharing to promote public trust', *Human Genetics*, vol. 137, no. 8, pp. 583–591.

Eckstein, L & Otlowski, M 2018, 'Strategies to guide the return of genomic research findings: an Australian perspective', *Bioethical Inquiry*, vol. 15, no. 3, pp. 403–415.

Evans, B 2018, 'HIPAA's individual right of access to genomic data: reconciling safety and civil rights', *Americal Journal of Human Genetics*, vol. 102, no. 1, pp. 5–10.

Frizzo-Barker, J, Chow-White, PA, Charters, A & Ha, D 2016, 'Genomic big data and privacy: challenges and opportunities for precision medicine', *Computer Supported Cooperative Work*, vol. 25, no. 2–3, pp. 115–136.

Genetic Information Nondiscrimination Act of 2008, 112 Stat. 881 (2008).

Hazel, JW & Slobogin, C 2018, 'Who knows what, and when?: a survey of the privacy policies proffered by U.S. direct-to-consumer genetic testing companies', *Cornell Journal of Law and Public Policy*, vol. 28, no. 35, pp. 35–66.

Health Information Technology for Economic and Clinical Health Act, under Title XIII of the American Recovery and Reinvestment Act of 2009, 123 Stat. 115 (2009).

Health Insurance Portability and Accountability Act of 1996, 110 Stat. 1936 (1996).

Jasanoff, S 2007, 'Making order: law and science in action', in E Hackett, O Amsterdamska, M Lynch & J Wajcman (eds), *The handbook of science and technology studies*, MIT Press, Cambridge.

Kichko, K, Marschall, P & Flessa, S 2016, 'Personalized medicine in the U.S. and germany: awareness, acceptance, use and preconditions for the wide implementation into the medical standard', *Journal of Personalized Medicine*, vol. 6, no. 2, E15.

Laurie, G 2014, 'Privacy and the right not to know: a plea for conceptual clarity', in R Chadwick, M Levitt & D Shickle (eds), *The right to know and the right not to know: genetic privacy and responsibility*, Cambridge University Press, Cambridge.

Lightbody, G, Haberland, V, Browne, F, Taggart, L, Zheng, H, Parkes, E & Blayney, J 2018, 'Review of applications of high-throughput sequencing in personalized medicine: barriers and facilitators of future progress in research and clinical application', *Briefings in Bioinformatics*, pp. 1–17.

Majumder, M 2018, 'United States: law and policy concerning transfer of genomic data to third countries', *Human Genetics*, vol. 137, no. 8, pp. 647–655.

Mendelson, D 2018, 'The European union general data protection regulation (EU 2016/679) and the Australian my health record scheme—a comparative study of consent to data processing provisions', *Journal of Law and Medicine*, vol. 26, no. 1, pp. 23–38.

Miller, A & Tucker, C 2017, 'Privacy protection, personalized medicine, and genetic testing', *Management Science*, vol. 64, no. 10, pp. 4648–4668.

Mostert, M, Bredenoord, A, van der Sloot, B & van Delden, J 2018, 'From privacy to data protection in the EU: implications for big data health research', *European Journal of Health Law*, vol. 25, no. 1, pp. 43–55.

National Institutes of Health 2019, *All of us research program*, viewed 1 April 2019, https://allofus.nih.gov.

Praisack, B 2017, *Personalized medicine: empowered patients in the 21st century?* New York University Press, New York.

Prince, AER 2018, 'Comparative perspectives: regulating insurer use of genetic information', *European Journal of Human Genetics*, vol. 27, no. 3, pp. 340–348.

Privacy Act 1988 (Cth).

Regulation (EU) 2016/679 of the European Parliament and of the Council of 27 April 2016 on the protection of natural persons with regard to the processing of personal data and on the free movement of such data, and repealing Directive 95/46/EC (General Data Protection Regulation) [2016] OJ L 119/1.

Salari, P & Larijani, B 2017, 'Ethical issues surrounding personalized medicine: a literature review', *Acta Medica Iranica*, vol. 55, no. 3, pp. 209–217.

Scolforo, M 2019, 'Genealogy database, Gum help convict DJ in '92 Killing', *Associated Press News*, 9 January 2019, viewed 1 April 2019, www.apnews.com.

Scudder, N, McNevin, D, Kelty, S, Walsh, S & Robertson, J 2018, 'Massively parallel sequencing and the emergence of forensic genomics: defining the policy and legal issues for law enforcement', *Science & Justice*, vol. 58, no. 2, pp. 153–158.

Shabani, M & Borry, P 2018, 'Rules for processing genetic data for research purposes in view of the new EU General Data Protection Regulation', *European Journal of Human Genetics*, vol. 26, no. 2, pp. 149–156.

Suthers, G, McCusker, E & Wake, SA 2011, 'Alerting genetic relatives to a risk of serious inherited disease without a patient's consent', *Medical Journal of Australia*, vol. 194, no. 8, pp. 385–386.

Wang, S, Jiang, X, Singh, S, Marmor, R, Bonomi, L, Fox, D, Dow, M & Ohno-Machado, L 2017, 'Genome privacy: challenges, technical approaches to mitigate risk, and ethical considerations in the United States', *Annals of the New York Academy of Sciences*, vol. 1387, no. 1, pp. 73–83.

White House, Office of the Press Secretary 2015, 'FACT SHEET: President Obama's precision medicine initiative', *Press Release*, 30 January 2015, viewed 1 April 2019, https://obamawhitehouse.archives.gov.

PART IV

The Future of Persons

CHAPTER 11

"The Obsolescence of Human Beings" and the Non-obsolescence of Law's Natural Persons: Transformations of Legal Personhood Through the Lens of "Promethean Shame"

Britta van Beers

Introduction

What does it mean to be human in a technology-driven society? The growing influence of movements such as transhumanism and posthumanism, both in academia and in popular culture, and the bestseller status of books proclaiming the end of humanity in various ways, such as *Homo Deus* (Harari 2016) and *Superintelligence* (Bostrom 2014), suggest that this complex issue resonates with a much wider public. Moreover, the question of what it means to be human is now also on the legal-political agenda. Striking examples are current debates on the legal recognition of self-learning robots as "electronic persons"; and

B. van Beers (✉)
VU University, Amsterdam, The Netherlands
e-mail: b.c.van.beers@vu.nl

M. de Leeuw and S. van Wichelen (eds.),
Personhood in the Age of Biolegality, Biolegalities,
https://doi.org/10.1007/978-3-030-27848-9_11

the legal-ethical controversies raised by the prospect of human germline editing and "CRISPR babies."

In this light, it is hard to believe that German philosopher Günther Anders' magnum opus *The Obsolescence of Human Beings 1: On the Soul in the Age of the Second Industrial Revolution* (Anders 2002a, author's translation) remains generally overlooked. Both this book from 1956 and its "sequel" from 1980 (Anders 2002b) offer a provocative philosophical anthropology for the technological age. The relevance of Anders' thoughts for contemporary discussions on technology already shines through his own description of the three central theses of *The Obsolescence of Human Beings.* These theses are:

> that we are no match for the perfection of our products; that we produce more than we can visualise and take responsibility for; and that we believe, that the things that we are able to do, we are also allowed to do, no: should do, no: must do. (Anders 2002a, p. VII, author's translation)

Anders' critical analysis of the impact of the increasingly technological lifeworld on our self-understanding as human beings and his reflection on the possibly dehumanizing effects of these developments, seem more relevant than ever. Indeed, many of the essays in the two books offer reflection on recent and upcoming technologies, such as human genetic engineering, artificial intelligence, and neuroscience. That Anders was able to foresee both these technological developments and the complex social-cultural dynamics to which they would give rise, is astounding, given that these technologies were, at the moment of his writing, still in an embryonic stage of development.

Also, in this respect, Anders' thought is related to the work of his close friends and fellow German-Jewish thinkers Hans Jonas and Hannah Arendt, with whom he studied under Heidegger, and fled from Germany to the United States during World War II. Arendt and Jonas similarly reflected on possible future technological developments. Moreover, like them, Anders' personal experiences during World War II profoundly influenced his philosophy. Indeed, the invention of the atomic bomb offers the most chilling example of the three main theses of *The Obsolescence of Human Beings.* Since Hiroshima we know that we are capable of bringing about the end of human history; yet, as Anders points out, we are incapable of imagining this nuclear apocalypse.

Nevertheless, unlike Jonas' and Arendt's philosophy, Anders' work has been largely neglected in the English-speaking world.

From this brief description, it will be clear that Anders' thoughts on the interaction between human beings and technological artifacts have proven to be remarkably applicable to current technological developments. The main argument of this chapter is that Anders' philosophical anthropology and, more specifically, his concept of "the obsolescence of human beings," is also highly relevant for current debates about legal personhood.

Throughout its history, the legal concept of the person has gone through several waves of naturalization and artificialization. At the moment, as will be discussed, a certain tendency can be detected to *artificialize* the legal concept of the *natural person*, that is, to "liberate the legal person" (Naffine 2011) from its naturalistic confines and humanistic premises. As a consequence, the gap between *homo* and *persona*, between flesh-and-blood human beings and legal persons, is widening. In this chapter I use Anders' essay *On Promethean Shame* as a lens to critically examine the current transformation of the legal concept of the natural person. The chapter's structure is as follows.

Hereafter, I introduce Anders' concept of Promethean shame, and identify three dimensions to this concept. As I discuss in the subsequent section, the idea of Promethean shame is able to throw new light on contemporary discussions about emerging technologies. I then argue that the three levels of Promethean shame can also be recognized within the current tendency to artificialize the legal concept of the person. Through a discussion of wrongful life claims, I highlight several problematic aspects of this more artificial approach. As I argue, the artificialist lawyer's radical neglect of the biological aspects of life leads to a distorted legal understanding of what is at stake within the legal regulation of emerging technologies.

From Promethean Pride to Promethean Shame

It is generally agreed that the myth of Prometheus symbolizes the human drive toward technological mastery over nature. In his essay *On Promethean Shame*, Günther Anders unfolds a radically different interpretation. "In a sense," he writes, "Prometheus' triumph has been all too overwhelming" (Anders 2016, p. 31). According to Anders, the modern Prometheus no longer experiences a sense of mastery and pride in the presence of his technological inventions. Instead, it is *shame* that he feels "when confronted by the 'humiliatingly' high quality of fabricated things" (Anders 2016, p. 30).

To substantiate his concept of Promethean shame, Anders commences his essay with an entry from his diary, in which he describes his first encounter with the phenomenon. Together with his friend T., he visited an exhibition of technology. From the start, T. acted very nervously. Each time that he viewed the highly complex appliances on show, he would lower his eyes, fall silent and conceal his hands behind his back, "as if he was ashamed to have brought these heavy, graceless and obsolete instruments into the company of machines working with such accuracy and refinement" (Anders 2016, p. 30).

T.'s odd pattern behavior toward the exhibited machines reminds Anders of the behavior of people who come eye-to-eye with their superior, or who encounter someone from a "higher" class, and suddenly feel embarrassed of who they are. Similarly, under the gaze of the shiny, seemingly perfect machines, T. is overcome by embarrassment about his own all too human origins. In Anders' striking words:

> *T. is ashamed about having naturally grown instead of having been made.* He is ashamed because he owes his existence to the blind and uncalculated, the highly archaic process of procreation and birth, which places him in stark contrast to the immaculate products, which are carefully designed through and through. His shame thus consists in his "*natum esse,*" in his lowly birth. He regards this is 'low' (not unlike the chronicler of founders of religion) precisely *because* it was a birth. Once he is ashamed of his antiquated line of descent, the same naturally also holds for its faulty and inescapable result: *he is ashamed of himself.* (Anders 2016, p. 5)

In the remainder of the essay Anders builds on this initial observation by offering further illustrations of the concept of Promethean shame and developing a theoretical framework to understand the phenomenon. A large part of his essay consists of responses to possible objections against his line of reasoning. One of the most obvious of these is why we would feel inferior to the things we make ourselves in the first place. In other words, what is the source of Promethean shame? Upon closer inspection, it appears that Anders identifies three interconnected sources for the phenomenon of Promethean shame.

The first source was already touched upon: T. feels ashamed that he has been *born, rather than made*. To those who argue that this shame is highly unlikely as it is humans themselves who make "their" machines, Anders retorts that the industrial production of artifacts, in practice, involves many makers. For assembly line workers and others who contribute to the production process, "the world of machines is no more an object of pride than it is 'their' property to own" (Anders 2016, p. 33).

This is all the more true for those on the consumer side of commodity production, whose agency and choices are an easy target for manipulation in today's market society. In reality, as Anders also writes, supply often leads to demand, and not the other way around (Anders 2016, p. 43). Indeed, in technological societies the real subjects of desire and demand are not so much people anymore, Anders writes, but machines. We have to give, so to say, the machine what it wants Therefore, "the maxim *become who you are* is [now] acknowledged to be the valid maxim of machines" (Anders 2016, p. 43).

This sense of human inferiority finds its source in a second layer of Promethean shame: the *shame for our given, embodied nature*. Compared to the speed, power, and accuracy of machines, we experience our own bodily "machinery" as rather primitive and lacking: "a faulty construction" (Anders 2016, p. 37). This sensation clearly surfaces in T.'s embarrassment for the plumpness of his hands. Unlike the artifacts we produce, our bodies are reshaped and prefixed, and, therefore, cannot live up to certain standards. More precisely, our bodies are not even raw materials: too "rigid and stubborn," they cannot be remade to fit the specification. The shiny world of products, in contrast, is one of incessant renewal and continuous innovation. In this light, the body emerges as "something that must be overcome" (Anders 2016, p. 36). Indeed,

> our body is today still yesterday's body, still the body of our parents, still the body of our ancestors. [...] It is morphologically constant; morally speaking: unfree, stubborn and unaccommodating. Seen from the perspective of machines, it is conservative, obsolete and un-changeable: a dead-weight in the rise of machines. In brief: we must invert the relation between freedom and unfreedom. *Things are free and human beings are unfree*. (Anders 2016, p. 38)

Our embodied nature also brings with it a structural vulnerability. This brings me to the third source of Promethean shame that Anders identifies: *our mortality and uniqueness.* In a way, the life of products is an immortal one. Evidently, objects at some point fall apart. However, when the object is replaced by a new one that is indistinguishable, it can be said to continue its existence through its successor. Anders likens this *serial existence* to "industrial re-incarnation" (Anders 2016, p. 51), a form of immortality achieved through the unlimited production of artifacts.

Human individuals, on the contrary, have to put up with their uniqueness: they cannot be replaced, and are thus marked by mortality. As such, "the experience and knowledge that we are not mass-produced [...] acts as a Memento mori" (Anders 2016, p. 56). This is what Anders calls "the human malaise of being unique" (Anders 2016, p. 51). In this light, the humanist credo, which holds that human beings are unique and therefore worthy of respect, has lost its meaning for "the transformed Prometheus." Instead, our uniqueness as human individuals become a burden of which we want to be relieved.

"Promethean Shame" in an Age of Trans- and Post-humanism

What to make of Anders' philosophical anthropology? Did Anders uncover a phenomenon that can be observed in daily life? Do we really experience, every now and again, feelings of humiliation, anguish, and defeat when faced with the blinding perfection of technological artifacts? Or is he exaggerating?

In the preface of *The Obsolescence of Human Beings*, Anders himself openly answers the last question in the affirmative (Anders 2002a, pp. 14–15). By using "philosophical exaggerations," that is, "misrepresentations pointing in the direction of the truth" (Anders 2002a, p. 86), he aims to awaken readers' imagination, thereby allowing them to open up for certain hidden or disquieting truths and enabling them to take a critical stance toward their growing dependency on machines and apparatuses. *On Promethean Shame* should, therefore, not be understood as a confirmation of human obsolescence in an age of technology, but rather as an *admonition*. Anders is concerned that technologies may result in "the annihilation of the human *as* human" (Anders 2016, p. 49). The paradox is that our Promethean aspirations to conquer nature, may also result in our own demise. From this perspective, Anders' negative anthropology can be compared to a black mirror: individuals who peer in this mirror, see their humanity reflected at its bleakest.

Even if Anders' methodology is one of the exaggeration, it is quite easy to recognize real-life instances of Promethean shame in contemporary discourse on new technologies. For example, upon reading Harari's *Homo Deus*, it is almost impossible not to be overwhelmed by feelings of human obsolescence in the light of technological developments to come. The Israelian historian predicts a fundamental devaluation of the human as a result of the rise of new technologies. As we come to rely on big data in all segments of life, the human will likewise dissolve in a sea of big data. Consequently, our humanist belief in human freedom and equality will be replaced by a *dataist* belief in the ability of artificial intelligence to predict human behavior through infallible, self-learning algorithms. Similarly, Anders' admonition that in technological societies the real subjects of desire and demand will be machines and not people, seems an apt description of already existing visions. For example, technology guru Kevin Kelly advises us to be smart and to listen to "What Technology Wants," as his 2010 book is entitled.

More generally, it can be observed that several authors tend to combine their awe for the technological possibilities to optimize the human condition with a disdain for the given, biological aspects of life. Especially for transhumanists, biology is something that must be overcome, as also expressed in the subtitle of Ray Kurzweil's transhumanist manifesto *The Singularity Is Near: When Humans Transcend Biology* (Kurzweil 2006).

Furthermore, also several scholars who are not self-acclaimed transhumanists are indirectly contributing to a collective sense of Promethean shame. A striking illustration is the book *The End of Sex and the Future of Human Reproduction*, written by law and technology expert Henry Greely. Greely predicts that within one or two decades, most prospective parents will reproduce artificially through a procedure that he calls "easy PGD": Preimplantation Genetic Diagnosis (PGD) with artificial gametes. As soon as gametes can be artificially produced through In Vitro Gametogenesis (IVG), PGD will become both very accessible and attractive, also to parents without fertility problems. Women would then no longer have to undergo hormonal treatment and surgery in order for their egg cells to be procured. Moreover, as soon as IVG is possible, parents will be able to produce hundreds of embryos without much pain or effort. They can then select the embryo with the "best" genetic profile for implantation. Greely thinks no rational person would say "no" to this technology:

> The process of human reproduction is wasteful, expensive, and bizarrely complicated. Such a process surely must be the product of evolution, because no one would have designed it this way. And, for many people who want to be parents, it does not work. Ultimately, this book is about the ways we are likely to redesign that system, to make it less wasteful, expensive and complicated. (Greely 2016, p. 44)

In other words, easy PGD will lead to a more efficient process of human reproduction. Natural reproduction, as "a product of evolution," has become outdated, according to Greely. From that perspective, human reproduction should be "redesigned."

Greely's words conjure up a world in which there is not much difference anymore between human reproduction and the production of commodities. From his perspective, human reproduction needs optimization to become more rational, transparent and cost-effective. If one goes along with Greely's line of argumentation, it is only logical if people in the future are ashamed of having been born the old-fashioned, clumsy way, rather than as the result of calculated and carefully designed methods of reproduction. It is even possible that this sense of shame, and its underlying social norm, will also find recognition on a legal level, though, for example, so-called *wrongful life* actions. In the next section I turn to the relevance of Promethean shame for current discussions about the legal person, using these wrongful life actions as a recurring example.

Legal "Metamorphoses of the Self"[1] in an Age of Promethean Shame

Pragmatist philosopher John Dewey once observed that the history of western culture shows a "chameleon-like change" of the legal person. Various legal concepts of the person have followed one after the other, yet "almost all concepts have persisted side by side in a confused intermixture" (Dewey 1926, p. 658). Nevertheless, a constant can be detected in this process. As I have discussed extensively elsewhere, the evolution of the legal person is, essentially, a dialectical one. The core question underlying this dialectical process is to what extent a legal person should be understood as the reflection of real-life, flesh-and-blood human beings, or rather as a disembodied and strictly legal-technical abstraction (Van Beers 2017, p. 569). The resulting dynamics can be described as a succession of alternating waves of *artificialization* and *naturalization*.

What is the main driver behind the continuous transformation of this legal notion? According to Dewey (1926, p. 655), the evolution of the concept has depended "on a mass of non-legal considerations: considerations popular, historical, political, moral, philosophical, metaphysical, and, in connection with the latter, theological." In these times, technological understandings of the human can be added to Dewey's list. In reaction to technological developments in the fields of artificial intelligence and biotechnology, a rather artificial concept is gaining ground in law. According to a growing number of legal scholars, the legal person is a fundamentally disembodied, "highly plastic" (Grear 2013), "mutable and fluid" (Naffine 2009, p. 45) entity. As they argue, the technological artificialization of the human, and the blurring boundaries between persons and things more generally, equally necessitate an artificialization of the legal concept of the person. They are no longer convinced by the classic legal distinction between so-called *natural persons* (human legal subjects) and *artificial persons* (non-human legal subjects such as corporations) in this technological, postmodern era. In other words, to them, the legal concept of the natural person has become obsolete.

In the remainder of this chapter I argue against the obsolescence of law's natural person by engaging with Anders' philosophy. Interestingly, the three sources of Promethean shame, which I identified in the previous section, are replicated on a legal level through three forms of contestation of the legal concept of the natural person. I will use wrongful life claims and their underlying concept of the natural person as a recurring example to illustrate these legal manifestations of Promethean shame.

The Shame of "Natum Esse"[2]

The legal importance of birth for recognition as a legal subject is probably best illustrated through Article 1 of the Universal Declaration of Human Rights: "all human beings are born free and equal in dignity and rights." Also in other fields of law birth is understood as an event of legal significance: when we are born in the physical world, we are simultaneously born as natural persons in the legal world.

However, certain asynchronicities between the biological birth of human beings and the legal birth of natural persons are emerging, especially within the legal governance of assisted reproductive technologies. A good example is the recognition of future children as legal subjects in these contexts, that is, children who have not been born or even conceived yet. The welfare of these future children has become one of the central principles in the law and ethics of assisted reproduction.

In several legal contexts, the discrepancy between legal birth and biological birth even takes on the shape of a clash. The current wave of lawsuits against fertility clinics suggests that birth is rapidly becoming the object of new legal contestations. If something goes wrong during a reproductive treatment, both the parents and resulting child can turn to tort law for recognition and compensation of their resulting grievances. For example, a doctor mistakenly selects and implants an embryo with a serious genetic disorder as part of preimplantation genetic diagnosis, or she wrongly assumes that there is no reason for prenatal testing, even if there is a family history of a certain genetic diseases. As a consequence of her culpable carelessness, the pregnancy is continued instead of terminated, resulting in the birth of a child with severe genetic disorders. Under these circumstances, both the parent and the resulting child may claim damages. These legal claims are better known as wrongful birth (the claim by the parents) and wrongful life actions (the claim by the child).

Within the logic underlying wrongful life actions, the shame to have been born takes on quite a literal meaning. According to, for example, the Dutch interpretation of wrongful life claims, the child is compensated for the fact of having been born instead of having been aborted. In the so-called *Baby Kelly* case,[3] the Dutch Supreme Court ("Hoge Raad") ruled that claimant Kelly's severely disabled life was so "pitiful," that damages for her entire life were deemed in order. An obstetrician had negligently failed to detect Kelly's grave genetic diseases during her mother's pregnancy. The Hoge Raad ruled that the obstetrician did not only negligently frustrate the parents' right to self-determination by robbing them of the possibility to terminate the pregnancy; it also ruled that the medical professional had violated her duty of care toward the future child.

This raises the question as to which of Kelly's legal interests were violated through the obstetrician's negligent failure. The damages awarded to Kelly do not leave much room for interpretation: Kelly received damages for not only the expenses related to her disabilities, but for all living expenses. By awarding her damages for what is essentially her life, the Supreme Court's ruling implies that Kelly had a legal interest in her own abortion. It is hard to deny that the Hoge Raad thereby makes a value judgment about Kelly's life: her life is unworthy of living and can therefore qualify as a source of damages. This position is hard to reconcile with the principle of human dignity (Van Beers 2018). This may explain why the legal construction chosen by the Hoge Raad is rather unique worldwide.

Furthermore, the recognition of a legal subject's interest in her own non-birth introduces a peculiar understanding of the legal person: a person who can legally and retroactively contest her own birth, which is, paradoxically, also the event that constitutes her legal subjectivity and allows her to act as claimant in the lawsuit in the first place. This radical separation between the legal subject and her birth suggests not only that, according to this logic, legal persons can preceed their own birth (this is, to a certain extent, already recognized through the well-known *nasciturus* fiction that dates back to Roman law). It also suggests that legal subjects can actively repudiate their own birth, thereby exhibiting a rather far-reaching form of *rejectio fati*. It is exactly this rejection of one's own fate that also seems to be at the root of the Promethean shame to have been born. Indeed, Anders (2016, p. 47) describes the attitude of "the transformed Prometheus" exactly in terms of *rejectio fati*.

The Body as "Something that Must Be Overcome"[4]

Also, the second source of Promethean shame—the shame for one's given, embodied nature—resurfaces in the context of current contestations of law's natural person. Once the notion of the legal person is understood as a purely artificial, legal-technical device, as an entity that it is entirely cut off from its real-life, flesh-and-blood counterpart, the human body can only be perceived as *ancillary* to the legal person. As a German legal scholar once described the consequence of his own purely legal-technical understanding of the person: "the embodiment of human beings is for their legal personality a completely irrelevant feature" (Zittelman 1873, p. 68).

If the human body can no longer be understood as an integral part of the legal subject (*persona*), only one possibility remains to represent the body in law's *summa divisio* of persons and things: it becomes an object of that subject's rights (*res*). This means that using a purely legal-technical, artificial concept of the legal person within the regulation of technological developments gives rise to a dualistic, cartesian understanding of the relation between body and person. The disembodiment of the legal person that results from this artificialist understanding of the legal person corresponds with a Promethean vision of the human body "as something that must be overcome" (Anders 2016, p. 36).

This tendency can be illustrated by a second, more common interpretation of wrongful life actions. Unlike the Dutch Supreme Court, most courts that recognize wrongful life claims award damages to children for only the expenses related to their disabilities. The well-known French case of Nicolas Perruche can serve as an example.

Nicolas' mother was infected with rubella when she became pregnant of Nicolas. However, her doctor negligently failed to detect this. The rubella resulted in Nicolas being born with Gregg syndrome, a congenital deficiency which manifests itself through multiple disabilities. If Nicolas' mother had known about the disease, she would have terminated the pregnancy. Both mother and child claimed they were therefore aggrieved by the doctor's negligent behavior and started a lawsuit for wrongful birth and wrongful life.

In 2000 the French Supreme Court (Cour de Cassation) recognized Nicolas' wrongful life claim.[5] However, unlike the Dutch Supreme Court, the court awarded damages for only the expenses related to Nicolas' disabilities. The decision led to much upheaval throughout French society. Also among legal scholars, the case was much criticized and discussed.

In deciding to hold the doctor liable for only the disabilities, the Cour de Cassation was able to evade making value judgments about Nicolas' life. However, as a result of this, another problem emerged. The disabilities that were regarded as the source of damages had not been caused by the doctor's negligence. Indeed, as Nicolas' disabilities were the result of genetic deficiencies, he could not have been born without them. The only other available scenario was that he would not have been born at all. In other words, interpreting wrongful life claims in this way stretches the concept of legal causation to its limit.

More precisely, one of its effects is to create a new legal fiction: the fiction that the child could have been born without disabilities. In this interpretation, the child does not have right or interest not to be born, as in the Dutch interpretation, but rather the right to be born in a different body. As this construction fictionalizes the way the natural person relates to his or her body, it could be said that the result is not just a legal fiction, but an entirely fictional legal subject, as Yan Thomas (2002, pp. 165–166) argues in his eloquent analysis of the *Perruche* case. As such, this new legal fiction takes the artificiality of law's natural person to a whole new level. By recognizing wrongful life claims, children are allowed to legally contest the natural state in which they were born, thereby dissociating their legal selves from their embodied selves in a radically new way.

Ultimately, the wrongful life claim makes it possible for a legal subject to complain about the fact of having been formed in the wrong shape, with the wrong body and the wrong genes. This brings to mind Anders' words "that the fact of 'having already been formed in the wrong shape' is the main defect that the human engineer seeks to address and the main reason for Promethean shame" (Anders 2016, p. 51).

The "Human Malaise of Being Unique"

The third and last level of Promethean shame takes root in the "human malaise of being unique" (Anders 2016, p. 51). As discussed, the modern Prometheus also feels inferior to the products he makes because he is, unlike them, irreplaceable. The so-called serial existence of things offers them, to a certain extent, an "immortality through reincarnation" (Anders 2016, p. 53).

Anders argues that this third source of Promethean shame goes back to a reversal of the humanist credo: human uniqueness is no longer to be regarded as a source of human dignity, but rather as a source of human inferiority in comparison to the endless production of things. Similarly, the humanist thought that "humankind is the measure of all things" seems to have lost all meaning for the modern Prometheus. Instead, in a world in which machines and products are perceived to outshine human beings in many respects, ***things will become the measure of all humans.***

Evidently, this reversal of the moral status of things and persons is not yet reflected on a legal level. Nevertheless, it cannot be denied that a dilution between the legal categories of persons and things is already becoming visible. Illustrations are recent proposals to confer legal personhood to entities that have traditionally been regarded as things, such as rivers, animals, and forms of artificial intelligence. In a similar vein, artificial persons such as corporations are increasingly successful in making themselves heard in a field of law that used to be the exclusive domain of natural persons: human rights law (Grear 2007, p. 511).

Also, in the context of wrongful life claims, a legal dedifferentiation between persons and things threatens to take shape. One of the concerns about wrongful life claims is that lawsuits like these introduce a new type of product liability, which portrays the child as the defective product in question. As Habermas writes about this type of lawsuit:

> [...] later generations can demand an account from the programmers of their genome; they can hold these producers responsible for what they, the offspring, consider the unwanted consequences of the organic starting point of their life histories. This new structure of attribution results from obliterating the boundary between persons and things. One can see this, for example, in the case of the parents of a handicapped child who hold their physician responsible for the material consequences of a mistaken prenatal diagnosis and undertake a civil suit for 'compensatory damages' – as though the medically unexpected handicap were tantamount to damage to one's property. (Habermas 2003, pp. 13–14)

Also, in other respects, the wrongful life claim can be used as an illustration of the Promethean shame for one's uniqueness. Within the Cour de Cassation's interpretation of wrongful life, the serial existence of products is mirrored in the legal fiction that the same individual could have been born in a different body without disabilities. Indeed, courts that award damages for expenses that are related to the child's disability counterfactually presuppose that if the medical professional's negligent behavior had not taken place, then a healthy version or copy of the child would have been born: a copy of the child without disabilities (Iacub 2002, p. 118).

The Non-obsolescence of Law's Natural Persons

I have argued that Anders' concept of Promethean shame is starting to materialize on a legal level through the current dilution between natural and artificial persons in law. In that process, technological possibilities to transform the biological aspects of life play a vital role. In this final section I take my analysis of legal personhood through the lens of Promethean shame one step further through a critical evaluation of this tendency.

Evidently, the outlines and meaning of the natural person, as a category of law and thus a human construct itself, are not set in stone. Concepts of the human and humanity have equally evolved over time. Without a doubt, technologies will lead to new answers to the question of what it means to be human. In that sense, law's natural person (*persona*), like its substrate (*homo*), is indeed "artificial by nature," to use Helmuth Plessner's famous expression. It should be noted, in that regard, that Anders himself expresses similar views of the human in a French article from 1937, in which he writes that "l'artificialité est la nature de l'homme et son essence est l'instabilité" (Stern/Anders 1937).

However, this artificiality cannot be the end of the discussion on the future of law's natural persons. The reason is that one need not be a transhumanist or posthumanist to understand how today's technologies make it possible to alter the *conditio humana*. Furthermore, one need not be a techno-pessimist to agree that these same technologies that may start of as liberating and empowering, can give rise to practices with disempowering and sometimes even dehumanizing effects. Current debates about big data, algorithms and the growing power of so-called *tech giants* provide ample examples.

Remarkably, Anders seems to have foreseen this dynamic of emerging technologies. Moreover, he underpins his critique of human self-instrumentalization with a concept of the person that is neither artificialist, nor naturalist or metaphysical. In a crucial section on the phenomenon of "human engineering," he explicitly denies that he aims to protect some version of human nature:

> Far be it from me, then, to regard the form (*eidos*) or morphological fixedness of existing 'species' (the 'human' species included) as 'sacrosanct' or 'good' just because they are what they are, or to retrospectively sanction them (or condemn their transformation as a 'cosmic prohibition'). All the more so, because nature itself is clearly open to mutation [...]. Even *if* there were such permanence, even *if* it were proven that the various species have endured unaltered to this day, this persistence would still not be morally conclusive, precisely because questions of what is just (*quaestiones iuris*) are independent of questions of fact (*questiones facti*). (Anders 2016, p. 48)

In his denial of being "metaphysically conservative" (Anders 2016, p. 47), Anders cannot be grouped with today's so-called bioconservative thinkers. Instead, what concerns him about human engineering is the danger that

> we are undergoing this transformation of the self for the sake of machines and because we employ these as blueprints and models for our own modification. We hence relinquish ourselves as our measure as humans and with this gesture we either restrict or give up our freedom. (Anders 2016, p. 47)

In other words, even if humans are artificial by nature, there are certain limits to their artificiality. These limits are reached as soon as humankind no longer functions as the measure of things, but instead ***things become the measure of humans.***

What are the implications of these thoughts for current discussion on law's natural person? Similar to the limits of human artificiality, there are also limits to the elasticity of law's concept of the natural person. The wrongful life claim illustrates how the risks of human self-degradation and self-objectification that Anders describes in the context of new technologies, can also manifest themselves on a legal level. The lives of children who are awarded damages for wrongful life, are measured according to a certain standard. This standard is one that is more commonly applied to defective things that are in need of replacement than to people.

More generally, artificial personhood, disembodied and empty as it is, cannot offer answers to the most vital legal challenges posed by technological developments. Ultimately, these technologies make answers to the question of what it means to be human indispensable, also on a legal level. Indeed, what seems to be at stake in the coming era of human enhancement technologies is our embodied, human nature. Artificial personhood, from that perspective, is a dead-end street. Without some concept of the natural person, the law cannot offer any guidance in these debates, and cannot counter the self-destructive potential of human engineering.

Notes

1. Anders (2016, p. 41).
2. Anders (2016, p. 30).
3. Hoge Raad, 18 March 2005 (*Baby Kelly*), *NJ* 2006, 606.

4. Anders (2016, p. 36).
5. Cour de Cassation, 17 November 2000, *JCP* 2000, II, 10438, p. 2293.

Bibliography

Anders, G 2002a, 'Die Antiquiertheit des Menschen 1', in *Über die Seele im Zeitalter der zweiten industriellen Revolution*, C.H. Beck, München.

Anders, G 2002b, 'Die Antiquiertheit des Menschen 2', in *Über die Zerstörung des Lebens im Zeitalter der dritten industriellen Revolution*, C.H. Beck, München.

Anders, G 2016, 'On Promethean Shame', in CJ Müller (ed.), *Prometheanism: technology, digital culture and human obsolescence*, Rowman & Littlefield, London, pp. 29–95.

Bostrom, N 2014, *Superintelligence, paths, dangers, strategies*, Oxford University Press, Oxford.

Dewey, J 1926, 'The historic background of corporate legal personality', *Yale Law Journal*, vol. 35, pp. 655–673.

Grear, A 2007, 'Challenging corporate 'humanity', legal disembodiment, embodiment and human rights', *Human Rights Law Review*, vol. 7, no. 3, pp. 511–543.

Grear, A 2013, 'Law's entities: complexity, plasticity and justice', *Jurisprudence*, vol. 4, no. 1, pp. 76–101.

Greely, H 2016, *The end of sex and the future of reproduction*, Harvard University Press, Cambridge.

Habermas, J 2003, *The future of human nature*, Polity Press, Cambridge.

Harari, YN 2016, *Homo Deus: a brief history of tomorrow*, Penguin, London.

Iacub, M 2002, *Penser les droits de la naissance*, Presses Universitaires de France, Paris.

Kelly, K 2010, *What technology wants*, Penguin, London.

Kurzweil, R 2006, *The singularity is near, when human transcend biology*, Penguin, London.

Naffine, N 2009, *Law's meaning of life, philosophy, religion, Darwin and the legal person*, Hart, Oxford.

Naffine, N 2011, 'Review essay: liberating the legal person', *Canadian Journal of Law and Society*, vol. 26, no. 1, pp. 193–203.

Stern (Anders), G 1937, '"Pathologie de la Liberté", Essai sur la Non-Identification', *Recherches philosophiques*, vol. VI, pp. 22–54.

Thomas, Y 2002, *Du droit de ne pas naître*, Gallimard, Paris.

Van Beers, BC 2017, 'The changing nature of law's natural person, the impact of emerging technologies on the legal concept of the person', *German Law Journal*, vol. 18, no. 3, pp. 559–594.

Van Beers, BC 2018, '"The best me I can possibly be", legal subjectivity self authorship and wrongful life actions in an age of genomic torts', in B van

Beers, S Sterckx & D Dickenson (eds), *Personalised medicine, individual choice and the common good*, Cambridge University Press, Cambridge, pp. 200–225.

Zittelman, E 1873, *Begriff und Wesen der sogenannten juristischen Personen*, Duncker & Humblot, Leipzig.

CHAPTER 12

Distributed Cognition, Distributed Being, and the Foundations of Law

Margaret Davies

Introduction

> …if we cancel the subject's claim to be first – the claim which surreptitiously keeps inspiring ontology – that which the schema of traditional philosophy calls secondary is no longer secondary either. (Adorno 1973, p. 139)

The human–nonhuman division runs through the idea and philosophy of law as understood in Western thinking. But it is only one way of understanding our place as human beings in the world. The division is, in fact, an anthropocentric mythology that obscures several important facts: first, that human life is part of and not fundamentally separate from life generally; second, that we are entirely dependent on the physical world and most particularly the land for our existence; third, that human beings are first and foremost physical beings. Scientific discourse traditionally draws taxonomic divisions between life forms and between life and the entire physical world, but also recognizes the continuities and co-dependencies between them. Appreciation of the connectedness, dependence, and

M. Davies (✉)
Flinders University, Adelaide, SA, Australia
e-mail: margaret.davies@flinders.edu.au

M. de Leeuw and S. van Wichelen (eds.),
Personhood in the Age of Biolegality, Biolegalities,
https://doi.org/10.1007/978-3-030-27848-9_12

physicality of human life is most thoroughly embedded in the biosciences, which throughout the past century have destabilized the foundation of Western ethics, philosophy, and law—that a human is a single unique organism with a natural character and pre-defined limits. As Gilbert et al. argue, "animals can no longer be considered individuals in any sense of classical biology: anatomical, developmental, physiological, immunological, genetic, or evolutionary" (2012, p. 334). Biotechnological innovations challenge our ontological architecture from a different angle, manufacturing and programming "nature" in a way that undermines its conceptual integrity as object to human subjects (Pottage 2007).

Instead of atomistic selves consisting of a binary mind/brain and body, the emerging model of the human is distributed and networked rather than centered. Cognition is a product of the "material engagement" of a plastic mind rather than located in a skull-encased brain (Malafouris 2013, 2015). At the same time, humanimal bodies are in constant exchange with a physical world understood as external but in fact integrated at all scales of human embodiment. The critical posthumanities and new materialisms have begun to adopt and further explore these scientific narratives of connectedness, reactivating decades-old feminist thought and in the process troubling not only the line between humans and nonhumans, but also the singularity of each human unit and its fundamental constitution. Much more effort, though still not enough, is being made to learn from non-Western philosophies, such as the many First Nations philosophies, that for millennia have understood the original connection of ourselves with each other and with the earth (Graham 2008; Kwaymullina and Kwaymullina 2010; Bawaka Country et al. 2016; Anker 2017).

Law has by and large continued to presume the disconnection of human from nonhuman, and to reproduce the image of individual natural human entities. That law is able to maintain such a position in the face of scientific evidence to the contrary may seem perverse but is based on two rather simple facts. First, modern Western law openly and necessarily constructs reality and has no need to *reflect* anything—hence law can deploy whatever outlandish fictions are instrumentally conducive to specific situations or contexts (if "outlandish" seems extreme, remember that a *ship* can legally be a *person*.) Second, while the foundations of law are ancient, it is a historically layered and culturally integrated artifact of multiple contexts and many geo-political crises and changes. Law has been formed by long-term reiterative processes, making it conceptually both complex and extremely resilient, albeit often incongruous with twenty-first-century knowledge.

Despite its constitutive and contextual nature, the located, constructed, *form* of law itself as a medium continues to be neutralized in Western thought. Faith in the neutrality of law as a system is illustrated by the fact that reform of particular laws is often regarded as a reasonable and sufficient response to failings in legal doctrine. This emphasis on reform is especially pertinent where some specific area of law, such as the regulation of biotechnology or definitions of causation or the understanding of harm can be redefined to respond to new knowledge. Over time, such interventions may accrue into deeper change. However, as critical legal theorists have argued for several decades, law reform in the traditional sense does little to address the underlying concepts of law that delimit what is legally possible. Therefore, it is not only the ways in which law addresses and regulates bioscientific interventions and discoveries that demands attention, it is also the structural components of law, or what Alain Pottage refers to as "legal form" (Pottage 2007). These are the broader conceptual aspects of law—property and intellectual property, the person, the family, modalities of governance, agency, identity, and so forth—that are often put to the test by bioscience, in particular its challenge to the nature and limits of the human being. In addition, however, there remains an even broader question about the form and imaginative possibilities of law itself—what we understand it to be in the first place. The dissolution of the separated human realm with its individuated beings puts the entire notion of law as the construct of human society under pressure. New ways of seeing law's conceptual structure become possible if normativity is understood not merely as the overt rules and principles laid down by courts and legislatures but as ingrained in material processes—including the materiality of human bodies in their physical contexts, but extending as well to nonhuman regularities.

How do we reorient the Western theory and concept of law as well as its constituent parts, to better account for these continuities? Some writers have commenced this project of reorientation but the challenges are considerable, given the institutional solidity of legal structures and their reliance on an imaginary of human separation (Grear 2011, 2015; Philippopoulos-Mihalopoulos 2011; Davies 2017). In considering the nature of law, it is important to consider not only its formal conceptual constitution, but also the cultural imaginaries and symbolisms that shape it. There are potentially several influential discourses informing the notion of law, but one that is particularly critical in the liberal, capitalist, and colonizing West has been, and continues to be, the imagery and conceptualization of property. The Western concept of property is a central tool for enforcing the separation between the human and the nonhuman spheres: it reflects and reinforces the human–nonhuman divide and supplies the imaginary upon which the conceptual framing of liberal law and its individual subject is based (Nedelsky 1990; Davies 1998). Broadly speaking, within the Western ontological and legal frame, nonhumans can be the objects of property while humans cannot be. At the same time, humans are the subjects of property, or owners, while nonhumans generally are not owners (there are, of course, many qualifications to be made to these very broad statements: see Naffine 2009; Davies and Naffine 2001). Just as significantly as the founding legal division between humans and nonhumans, when reified as legal person, the human being is imagined through a proprietary metaphor as self-owning—bounded and self-possessed (MacPherson 1964). It is impossible therefore to understand the idea of the legal person without understanding both its conceptual *opposition* to property and its *reliance* on the imaginary of property as autonomous and largely exclusionary.

This chapter has two parts. First, I examine the constitution of the Western version of law through the lens of the differentiation of the human from the nonhuman, a distinction that is reified (though not perfectly) in the concepts of person and property. Second, I consider what might be the possibilities for a view of law that does not so thoroughly rely upon this division. Even articulating these objectives is daunting, since law derived from the civil and common law traditions of Western Europe is so thoroughly grounded in and imbued with the human–nonhuman distinction, its singularized persons and objectified property.

The Order of Life and Law

Law is intrinsically social, but what, to begin with, does this mean? Until recently the presumed social basis for law was essentially human. But the definition of the 'social' has now moved beyond exclusively human relationships, providing new challenges and opportunities for legal theory (Latour 2005, p. 6; Bennett 2010). The socio-legal sphere is undoubtedly material, but the "social" and the "material" have gained an exceptionally expansive meaning in humanities and social science research, drawing in flesh and bodily relationships (which are inevitably also ecological), actions and relationships, everyday practice, and particular spaces (Beasley and Bacchi 2007; Ewick and Silbey 1998;

Cooper 2014; Delaney 2010). We are seeing a greater attention to the physical nature of the social and to the fact that humans are formed as relating entities located in time and place/space (Bawaka Country et al. 2016). The biological organism of the human is no seen as longer one thing, or as especially separate from an environment. Our individuality, sense of self, boundedness, mentality, and all of the cultural forms associated with ourselves as humans are now understood to be the effects of relationships. These relationships are, in the first place, (chronologically and ontologically) ecological and physical, subsequently social and political, and finally singular and individual. They are relationships in which both humans and nonhumans play an active role: rather than being the passive recipient of human action, nonhuman life and matter plays its own part in creating a social life in particular places.

A first consideration therefore relates to the onto-epistemology of law, persons, and property and their constituent parts (in particular subjects and objects). Barad's term "onto-epistemology" denotes that being and knowing are not separate and that we bring things into being as known objects and entities (Barad 2007). This is not to deny the independent existence of the world or to suggest that we can know it in any way whatsoever. The term applies to everything, but is particularly apt for recognized artifacts like property, the legal person, and law, where there is never an essential thing to describe that is independent of how it is understood and defended (Leiter 2011, pp. 669–670).

Which comes first in an onto-epistemological sense, law, persons, property or the human–nonhuman division? Until recently in Western legal theory it may not have made much sense even to ask this question. Clearly the human–nonhuman division has been seen as "natural" and ontologically prior to law and society, while the legal constructs of "person", "subject", "property", and "object" are regarded as artificial and secondary constructs of this perceived structure. But the order of things (causal, temporal, and taxonomic) is disrupted if we question the priority of the human–nonhuman divide and instead understand it not as a first cause upon which everything else is based, but as a construct that emerges with our (Western) normative ordering of the world, including our law. If humans are distributed through networks of being and knowing, if symbiosis is more definitive of biological existence than individualism (Gilbert et al. 2012), if cognition is reliant as much on extended relations as on internal processes (Malafouris 2015), then the "human" as a material-cognitive entity is only the secondary product of these networks. In this re-ordered account, property, the person, and law might be understood as co-emerging with the human–nonhuman divide as we understand it: legal concepts are not the effect of a prior division of human from nonhuman, but rather an integral part of a matrix of fixations that produce and affirm this division. These fixations can be variously described but cross the span of Western social life and are cultural, political, legal, economic, and religious. Property, the legal person, and law reflect a presumed division between human and nonhuman but are also key mechanisms for producing and maintaining this distinction.

One traditional way of thinking about our (Western, Anglo-centric) law is that it is a construct within a construct. As we understand it in the West positive law is normally associated with a state and is conventionally regarded as somewhat autonomous from social life generally: it has distinct institutions, forms, procedures, bureaucracies, and officials that set "law" apart from other socially constructed norms. From another perspective (but still within the Western mindset), "law" is not at all autonomous from social life as its entire existence is coextensive with, and inseparable from, human action.[1] But regardless of the position taken on whether law is relatively separate from or entirely enmeshed with social norms, state-based positive law can nonetheless be regarded as the ethnocentric product of a particular world view—Western humanist culture and its anthropocentric philosophy—which is itself a construct in the sense that it has emerged from specific, located, and strongly shared cultural histories. That world view is anthropocentric—though it was born in theocentrism—and separates human life from essentially everything else. Thus, the construct of positive law sits within the constructs of Western theory and philosophy based on a fundamental division of humanity from everything else. In a broad but not entirely consistent fashion, this division of subject from object leaves human beings free to dominate and appropriate the nonhuman world of objects.

This narrative, however, may underestimate the extent to which property, persons, and law have themselves been responsible for solidifying and entrenching anthropocentrism and the separated self (see generally Nedelsky 1990). In his refutation of the idea of natural rights and in particular the purportedly natural right of property, Jeremy Bentham famously observed that "Property and law are born together and die together. Before laws were made, there was no property; take away laws, and property ceases" (Bentham 1931, p. 113). Bentham's statement is a simple recognition that all property is a product of positive law, and does not pre-exist it. Property is secondary to law and exists entirely within it. The statement can, however, also be extended to the acknowledgment that the form of law and the form of property are intermingled: not only is positive law the origin and guarantor of property but property supplies the imaginary forms—terrain, boundedness, exclusiveness, hierarchical power—for state-based law (Davies 1998). Even further, and to modify the aphorism, we could say that "property and law and the person are born together and die together". Scholarship has for some decades explored the connections between the idea of the autonomous self or "possessive individual", the imaginary of property, and the concept of law, as well as the overlaps between the categories of person and property (see for example, MacPherson 1964; Nedelsky 1990; Frow 1995; Naffine 1998; Davies and Naffine 2001). The scholarship underlines the point that the law, the person, and property are not distinct but form a tightly conceptualized triad, linked in form and structure, existentially interdependent, and imagined through similar tropes. Each node has a separate role and is regarded as autonomous from its other two terms, but in fact law, persons, and property are linked by both technical and symbolic means (Davies 1998). Importantly for my point here, the triad is complicit in maintaining the boundaries beyond positive law between human beings and our nonhuman others. The human actor is reified and protected (as a racialized and gendered ideal type of man) as self-proprietor while the nonhuman world becomes the passive object of law and property.

An example of these dynamics of separation can be seen in the displacement of law in colonialism. The so-called "common" law of present-day Australia exists here because it was removed from its own indigenous conditions in England and transplanted across the globe to a quite different location, resulting in the Aboriginal laws of the Australian continent being obscured and suppressed. In this transplantation, as Nicole Graham argues, English "common" law, as well as its property law, is dephysicalized because it is perceived as having no intrinsic connection to its physical context and in particular the land (Graham 2011). Law is seen as entirely positive or imposed by a sovereign (and I note here the connection between imperium/sovereignty and dominium/property: Cohen 1927). This is not to say that Anglo-Australian law is entirely unresponsive to local conditions or has not been adapted over the past 230 years: in fact, once transported, Australian colonial law diverged from British law in many ways. None of these divergences were, however, fundamental to the structure and nature of the law itself. The "skeleton of principle",[2] to use a notorious phrase, remains intact as does the conceptual shape of the law and its abstract account of property in which land and things are essentially fungible. Thus, (Western) law acts upon the people, land, and physical environment but is not intrinsically connected to them or seen as part of them. Rather, Anglo-Australian positive law seems to float above human society and above our physical context in a mainly ideational space.[3]

The transportation of law from one place to another is only imaginable under certain conditions, essentially that it is understood as a human artifice, that can be separated as an objective thing from place and society, and is universal in form though not in content. The portability of law requires that the substance of law be seen as entirely a human construct rather than universal as it is in the natural law tradition: in other words, mobile law is a creation that can be picked up (as it were) and taken to a new place. Put another way, being taken from one place to another means the law did not pre-exist in its new location but had to be imposed there. It is consistent with this idea of a movable positive law that there could nonetheless exist across the globe a universal "natural" law: classical natural law writers envisaged positive (for example, local) law co-existing with natural law. Both the concepts of natural and positive law, however, separate law from physical place and particular societies: the natural law is separate precisely because it is grounded nowhere and everywhere (it does not have a place) and the positive law is separate conceptually because it can be moved from place to place.

The fact that the transportation of law across the globe is imaginable means law as a concept is not intrinsically connected to the place it is removed from or the bodies that made it and perform it in their everyday lives. In order for this to be possible, such a notion of law must be understood as essentially abstractable from its own context and society, that is, as an ideational rather than a material form. It is true that early liberal writers such as William Blackstone attached the law to embodied subjects: law went with the (human) person and English subjects (like Roman citizens) took their law with them—at least when they went to places where there was no pre-existing law (or so they thought): "if an uninhabited country be discovered and planted by English subjects, all the English laws then in being, *which are the birthright of every subject*, are immediately there in force" (Blackstone 1765, pp. 104–105, emphasis added). With some exceptions concerning the relevance of the law in question, a single Englishman on an uninhabited island brings his law with him.

But does he? Or rather, can he? The idea that law is conceptually detachable from place, from the entire nonhuman world, and even from the located human society that created it, is only one story about law. It is a story in which dominant humans—such as Englishmen—have the ability and the right to take themselves and therefore their law all over the globe. In this view law, like human rights, can be attached to persons but not to places. This concept of law connects it to the person, presumes the division between human and nonhuman, and perpetuates an image of law which is a construct within an entirely human context. It reduces the nonhuman world to an other and promotes and enables dominance of this other through concepts such as property. Indeed, in the mindset where humans are originally separate from nonhuman things, all of the others (and even some of the humans) can potentially be appropriated.

Post-Blackstone, the idea of law became even more fully detached from its context: it was separated in positivist theory from humans as well as from places and societies and identified with the sovereign, the state, or some other ideational foundation. There is no doubt that the notion of institutional positive law has a self-fulfilling meaning as a legal imaginary or reified image of law. It does exist and is correct as a description of Western law. Broadly speaking, statism and ideals of national unity led to a notion of law that aimed to consolidate, smooth out, and unify previously differentiated local customary laws. Similarly, positivist legal theory, the theory that insisted that law was created by human society as a special and separate set of institutionalized norms, removed the concept of law from the people and from their everyday lives. It is a powerful narrative that shapes the Western world and cultures subjected to it, but does it capture the reality of Western law? Is it true?

Looked at another way, it is possible to set aside this narrative of human exceptionalism and its separate law. There are other possible narratives and other stories about law. Some of these alternative legal mythologies are historically submerged and somewhat forgotten while others have been emerging in theories of the past century. These counter-narratives challenge both the notion that law, including Western law, is essentially abstract and the idea that its materiality is entirely human. A long historical review would, for instance, reveal pre-national European ideas and practices of law that were often localized and specific to the nexus of human society and place. Roman law was, originally, the law of a small city state (Ibbetson and Lewis 1994, p. 11). Customary laws were understood as local and specific—referring to *this* forest, *this* tree, *these* people, *this* distribution of resources, *these* rituals and relationships.

In the twentieth century, the positivist detachment of law from society, persons, and place was challenged by expanding the notion of law, by observing its material basis, and by critiquing the fiction of separation. Legal pluralist, as well as realist, feminist, minority, and critical legal theories of the twentieth century depicted the statist and positivist notion of law as only one form of law, which was not as dephysicalized or separated from politics and the everyday as its theorists supposed. The entire twentieth century of Western legal theory can perhaps be seen as encompassing at once the pinnacle of legal positivist thought as well as multiple counter-narratives emphasizing law's entanglement with the social and political domains.

Thinking Life and Law Together

Thinking of the human condition and therefore the social sphere as fully connected to the physical and ecological world seems obvious. Perhaps the point about our Western theoretical past is not that it did not comprehend this obvious point but rather that through a series of philosophical, religious, and socio-cultural interventions it cut the available material in such a way as to over-emphasize, and ultimately to conceptually segregate, a distinctively human sphere. The problem of course is that everything else, the entire nonhuman world (most of it) is thereby forgotten, if not entirely foreclosed, from theory and reduced to object status. Moreover, the distinctive moves by which this exclusion was achieved created and reinforced time and again a racist and gendered division in thought itself (Plumwood 1993; Mills 1997; Vial 2016). These moves include the separation of mind from body, the insistence that animals were nothing more than machines driven entirely by instinct, the division of peoples into civilized and uncivilized, the gendering and racializing of culture and nature, and endless other divisive and exclusive notions designed to promote the supremacy of the European male (Plumwood 1993, pp. 105–109). While the human–nonhuman distinction is clearly a critical axis of the tradition of Western philosophy, as many critics have noted, the several other distinctions with which it is partially aligned have the effect of separating out a particular being—rational (white) man—as the central, most human, example of the human (Merchant 1980; Lloyd 1984). Everybody else has been left with the struggle to prove themselves equal while the built-in privilege of the benchmark status is culturally disguised as merit.

The late twentieth and early twenty-first centuries have generated an explosion of theory that challenges these separations—both the individualized human with its divided mind and body and the distinction between the human and the nonhuman. To return to the question I posed at the outset, how can this theory be mobilized to reimagine law? As I have explained, it is not only the content of law (its doctrines and concepts) that are implicated in the constitution of a delimited and binary world, but also the form of law, the concept of law itself.

As I have suggested, the reimagination of law as such is an extensive and difficult project. It cannot be driven by simple theoretical interventions, but will rather emerge from incremental, collective, and accumulated change over a long period of time. Onto-epistemological categories are resistant to change, precisely because they govern symbolic as well as socio-political systems. Nonetheless, theory can offer imaginative syntheses and extensions of present knowledge and there are two angles that I would like to mention briefly by way of promoting the reimagination of law. In this context I cannot develop them fully, but they do, I hope, provide some inkling of the possibilities for rethinking law at the conceptual level. The first issue specifically concerns traditional narratives of property that connect it to a narrow form of exclusively human labor and the second looks at the nature of normativity as an embodied and material rather than an ideational category.

Property and Agency

John Locke's description of and justification for private property is very well known and has been extensively discussed and critiqued. Within the colonial territories of the Anglosphere it is undoubtedly one of the more powerful justifications for property because it provided a rationale for turning occupation into ownership and sovereignty. Having (falsely) characterized the lands to be taken as previously unoccupied or at least unowned because they appeared to be uncultivated (Tully 1993, pp. 137–176), the apologists for colonialism viewed the mixing of the colonists' self-owned labor with the raw materials of the natural world as a normative basis for ownership. The argument assumes self-ownership of European people in a context of seeming natural abundance: it was calculated to underpin colonial expansion to areas of the world whose social arrangements did not reflect European norms of individual autonomy and whose cultivation practices were willfully misunderstood (Hsueh 2006). It continues to resonate, at times very strongly, in arguments that certain resources are "under-utilized" (for instance by present-day Aboriginal custodians) and should be open for economic exploitation.

Lockean justifications for private property are based—like the private property system in general—on a division between the intentional laboring creator, who becomes an owner, and the natural worked-upon material, that becomes the owned object of property. The division is both produced and reaffirmed by the justification: labor is specific and unidirectional (man fences unoccupied land and grows crops) and often involves a conflict against, or at least an exertion of will over, the natural other. Hence, the labor of human reproduction is not accorded political significance. Indeed *as labor*, reproduction in Locke is entirely appropriated and controlled by men "in order to ensure certainty of paternity, and, hence, certainty of inheritance" (Clark 1977, p. 722). The labor of First Nations peoples was similarly unrecognized as sufficiently appropriative to lay the foundation for property: many Indigenous peoples, after all, worked with rather than against the natural world and did not subdue it in the biblical sense necessary for Lockean appropriation (Pascoe 2014).

The concept "man fences unoccupied land and grows crop" tells a very narrow story about labor which may have been plausible to a certain extent in the seventeenth century, but is hardly appropriate today. In focusing upon the conscious and intentional struggle of the individual white man, and like many anthropocentric narratives, the labor justification not only leaves much space for the marginalization of many humans, but also reduces the biophysical activities of the earth to "the spontaneous hand of nature" (Locke 1988, p. 286): the production of water, the evolution of plants, the growth and maintenance of soil fungi and nutrients, the chemical formation of the atmosphere, let alone the long geological processes that have preceded and surround life, are all foreclosed from the European narrative of man's creative labor. Hence, and only because of this foreclosure, the benefits and the normative attachments in the form of property all accrue in this story to the white man.

The essentially distributed and situated nature of being, and of the forms of labor that maintain being, is sidelined in accounts that focus on individual intentional productive labor. There might be many alternative starting points for a reconfigured account of property, but in my view there is also merit in repurposing traditional accounts—already ingrained in Western thought—in such a way that they become imbued with new thinking. Locke-inspired accounts of property are culturally powerful: can this influence be strategically deployed? One possibility is to retain the emphasis on labor, production, reproduction, and generativity and see where that leads in normative terms. The foundation, however, must be that generativity is not (and can never be) based solely on individual labor or even human labor, but is cyclical and is comprised of distributed co-production between human and earth. There is never any physical thing created by human beings that has not already, and for billions of years, been generated and that continues to be shaped by geological and biological processes that are far beyond any present human capacity.

This does not necessarily mean that the human input should not be accorded some legal significance, but that it needs to be seen as only one element of the normative outflow of generative processes and not unduly rewarded to the detriment of other participants. What other normative consequences exist? First, a rather obvious normative consequence of thinking of all human production as co-production is that the human producer owes a debt, to use rather conventional legal language, to the earth for the product: there is therefore an obligation, which may be framed in Lockean terms as the obligation to leave "as much and as good" for the earth itself (rather than for future human exploiters). Even within existing legal paradigms, such obligations can potentially be accorded legal value through notions such as stewardship and custodianship (Lucy and Mitchell 1996). A different position would accord personality and even rights to the nonhuman entity itself, either because it has a subjectivity that ought to be recognized by law or because it is simply a practical way to raise its profile and ensure that it has legal protection (Stone 1972; but regarding practical efficacy see counter arguments made by O'Donnell 2019). Transforming all actants or creators, or even all entities of value (animals, rivers, etc.) into "subjects" becomes very difficult to conceptualize: unless we are subjects without objects or dynamically circulating subject-object entities, turning nonhuman entities into subjects may simply be "an exercise in boundary extension, which otherwise retains the basic conceptual framework" of self-other (Plumwood 2002, p. 143). This is not to deny entirely the idea that a nonhuman entity could be attributed some form of personality or subjectivity, but it is a concept that needs to be approached with caution. Finally, and more promisingly, understanding human–nonhuman generativity as a point of departure for a normative reconstruction of property may supplement an emerging view of property as place-based, "a connection which acknowledges both place agency and the agency of people in place" (Bartel and Graham 2016, p. 278). A place-based view of law and property would understand norms as emergent from place-human connections rather than emanating from an abstract system which is indiscriminately applied to various places.

Embodied Norms

Trying to extract normative and potentially legal forms from a factual foundation begs several significant questions: what *are* norms/laws and how do they relate to the material facts of the world? As I have suggested, the form of law as understood in its Western forms has become extremely constraining: some issues can be addressed by changing the substance or content of law but such a strategy is always limited by the perceived nature of law, which is based entirely on a division between human and nonhuman. Traditionally in legal and political philosophy, norms and facts have been regarded as entirely separate: norms are ideational and prescriptive, while facts are material. Norms are seen to govern and construct facts: they are acts of human will that shape the (passive) material world, both human and nonhuman. Positivist legal theory entirely adopted Hume's prohibition against deriving norms from facts, and firmly defended this line for 200 years (Hume 1969, p. 521). In positivist legal theory, the "is"—the world of material fact—is sharply delineated from the legal "ought"—an act of human will channel through legal institutions (Austin 1954, p. 184). Norms of indeterminate status, such as social or so-called "natural" morality, were essentially irrelevant in the strong version of this schema.

The norm/fact distinction has an instrumental relevance in a world that continues to rely on it. Over time, however, the architecture of law changes, and it is possible to prefigure alternative forms that might emerge in the future. In a world where the distinctions (mind-body, human-nonhuman, culture-nature, and so forth) that have informed the fact-law separation are increasingly untenable, it is reasonable to suppose that law will also be understood differently in the future (see for example, Davies 2017). There are now extensive theoretical resources that have been accumulating over at least 50 years that critique the givenness of the norm-fact distinction and which deploy a language that bypasses or reconfigures it (see for example, Merchant 1980, p. 4). Any "factual" description, for instance, carries with it the ingrained normative assumptions of words and theories. Repeated facts become regularities and conventions that are, or can turn into, norms (Lewis 1969; Gilbert 2008; Stolzenberg 2010). The concept of performativity proceeds from the perception that social and legal norms come into being through repeated embodied actions (see for example, Butler 1990; Blomley 2013; Davies 2017). On the other side of our conventional divide, nonhuman life, including plants, are increasingly understood to possess a number of the qualities previously reserved for, and understood as characteristic of, humans (on the ability of plants to learn see Gagliano 2018, pp. 60–64). Does this therefore mean that they are capable of not just automatic but also agentic norm-following behavior?

An extension of ideas regarding normativity (and by association legality) situated in the co-becoming of human and nonhuman can be imagined. Indeed, as soon as normativity is recognized as embodied—because it relies on material action including the physical processes of signification—the path is laid for a view of normativity and law that is ontologically extended and distributed, like human beings. It is here that potential exists for a bioscientifically informed theoretical approach to law. In this theoretical space, there is no room for an idea of law as only a human artifact, since such a thing would be impossible. The human is never *only* human. As indicated above, the idea of law as a human construct only now exists as a result of the foreclosure of physical existence from descriptions of human law. Potentially, the idea would be for a view of law interconnected with ecosocial and biophysical relations: neither "natural" nor entirely human, but formed in the emergence of the human from an extended materiality. Such an account could give rise to an understanding of *biolaw* which is neither the application of law to bioscience nor an account of the fundamental laws of biology, but rather the generative and regenerative formations of both law and life: the patterns, iterations, symbiotic relations, responses, and behaviors of life understood as law.

Conclusion

"Law" as a positive human construct has usually been differentiated from the natural laws of science.[4] Human law is created; laws of nature are part of the fabric of the universe. Human laws can be changed at will; the laws of nature can only be discovered and refined in human understanding. Human laws can be directly transgressed; the laws of nature cannot be. One of the culturally troubling aspects of bioscientific and biotechnological interventions is that they seem to operate in the zone between the human and the nonhuman and therefore challenge this perceived dividing line between things that can be changed and things that cannot be. At first, biotechnological interventions in life appeared illegitimate. But perhaps the distinction between laws of nature and human laws was never entirely clear. Perhaps at one end of a spectrum there *are* immutable physical laws of the universe, as best described by science, and at the other end there are the entirely mutable laws, regulations, and social norms of human associations such as nation states and cultural groups. In between, are material regularities and patterns that are normative in the sense that they mandate the behaviors of physical objects, of life forms, and of humans but are of varying degrees of mutability and immutability. It is possible that conceptual connections between these different forms of law or norms can be described which bring together not only bioscience and law, but also the physical sciences and the social sciences generally.

Notes

1. The entire twentieth century in legal theory can be read as a struggle between theorists who argued for some kind of separateness and unity of the concept of law, and those who saw it as entirely enmeshed with social life and politics, and pluralistic rather than unified. Part, but not all, of the debate is explicable by reference to choices of perspective. For my own view, see Davies *Law Unlimited: Materialism, Pluralism, and Legal Theory* (Routledge, 2017).
2. The term 'skeleton of principle' was used by Brennan J in *Mabo* to denote the limit of the court's power in responding to history. See *Mabo v Queensland* [No 2] (1992) 175 *Commonwealth Law Reports* 1, 29.
3. Although the notion was rejected as a description over a century ago, there is a sense in which the idea of positive law remains a 'brooding omnipresence in the sky': Justice Oliver Wendell Holmes in *Southern Pacific Co v*

Jensen (1916) 244 US 205, 222. Holmes rejected the idea that law was a 'brooding omnipresence' in favour of seeing it as 'the articulate voice of some sovereign'.

4. Positive human law is also frequently differentiated from so-called 'natural' law—the law that is said to emanate from human nature or from god.

References

Austin, J 1954, *The province of jurisprudence determined*, Weidenfeld and Nicolson, London.

Adorno, T 1973, *Negative dialectics*, Continuum, New York.

Anker, K 2017, 'Law as … forest: eco-logic, stories and spirits in indigenous jurisprudence', *Law Text Culture*, vol. 21, pp. 191–213.

Barad, K 2007, *Meeting the universe halfway: quantum physics and the entanglement of matter and meaning*, Duke University Press, Durham.

Bartel, R & Graham, N 2016, 'Property and place attachment: a legal geographical analysis of biodiversity law reform in New South Wales', *Geographical Research*, vol. 54, no. 3, pp. 267–284.

Bawaka, C et al. 2016, 'Co-becoming Bawaka: towards a relational understanding of place/space', *Progress in Human Geography*, vol. 40, no. 4, pp. 455–475.

Beasley, C & Bacchi, C 2007, 'Envisaging a new politics for an ethical future: beyond trust, care and generosity—towards an ethic of "social flesh"', *Feminist Theory*, vol. 8, pp. 279–298.

Bennett, J 2010, *Vibrant matter: a political ecology of things*, Duke University Press, Durham.

Bentham, J 1931, *Theory of legislation*, K Paul, Trench, Trubner, London.

Blackstone, W 1765, *Commentaries on the Laws of England, Vol 1*, Oxford Clarendon Press.

Blomley, N 2013, 'Performing property: making the world', *Canadian Journal of Law and Jurisprudence*, vol. 26, pp. 23–48.

Butler, J 1990, *Gender trouble: feminism and the subversion of identity*, Routledge, New York.

Clark, L 1977, 'Women and John Locke; or, who owns the apples in the Garden of Eden?' *Canadian Journal of Philosophy*, vol. 7, no. 4, pp. 699–724.

Cohen, M 1927, 'Property and sovereignty', *Cornell Law Quarterly*, vol. 13, pp. 8–30.

Cooper, D 2014, *Everyday utopias: the conceptual life of promising spaces*, Duke University Press, Durham.

Davies, M 1998, 'The proper: discourses of purity', *Law and Critique*, vol. 9, pp. 147–173.

Davies, M & Naffine, N 2001, *Are persons property? legal debates about property and personality*, Ashgate, Farnham.

Davies, M 2017, *Law unlimited: materialism, pluralism, and legal theory*, Routledge, Abingdon.

Delaney, D 2010, *The spatial, the legal, and the pragmatics of world-making: nomospheric investigations*, Routledge, Abingdon.

Ewick, P & Silbey, S 1998, *The common place of law*, University of Chicago Press, Chicago.

Frow, J 1995, 'Elvis' fame: the commodity form and the form of the person', *Cardozo Studies in Law and Literature*, vol. 7, pp. 131–171.

Gagliano, M 2018, *Thus spoke the plant: a remarkable journey of groundbreaking scientific discoveries and personal encounters with plants*, North Atlantic Books, Berkeley, CA.

Gilbert, M 2008, 'Social convention revisited', *Topoi*, vol. 27, pp. 5–16.

Gilbert, S, Sapp, J & Tauber, A 2012, 'A symbiotic view of life: we have never been individuals', *Quarterly Review of Biology*, vol. 87, no. 4, pp. 325–341.

Graham, M 2008, 'Some thoughts about the philosophical underpinnings of Aboriginal worldviews', *Australian Humanities Review*, vol. 45, pp. 181–194.

Graham, N 2011, *Lawscape: property, environment, law*, Routledge, Abingdon.

Grear, Anna 2011, 'The vulnerable living order: human rights and the environment in a critical and philosophical perspective', *Journal of Human Rights and the Environment*, vol. 2, pp. 23–44.

Grear, A 2015, 'Deconstructing anthropos: a critical legal reflection on "anthropocentric" law and anthopocene "humanity"', *Law and Critique*, vol. 26, pp. 225–249.

Hsueh, V 2006, 'Cultivating and challenging the common: Lockean property, indigenous traditionalisms, and the problem of exclusion', *Contemporary Political Theory*, vol. 5, pp. 193–214.

Hume, D 1969, *A treatise of human nature*, Penguin, Middlesex.

Ibbetson, D & Lewis, A 1994, 'The Roman law tradition', in A Lewis & D Ibbetson (eds), *The Roman law tradition*, Cambridge University Press, Cambridge.

Kwaymullina, A & Kwaymullina, B 2010, 'Learning to read the signs: law in an indigenous reality', *Journal of Australian Studies*, vol. 34, no. 2, pp. 195–208.

Latour, B 2005, *Reassembling the social: an introduction to Actor-Network Theory*, Oxford University Press, Oxford.

Leiter, B 2011, 'The demarcation problem in jurisprudence: a new case for scepticism', *Oxford Journal of Legal Studies*, vol. 31, pp. 663–677.

Lewis, D 1969, *Convention: a philosophical study*, Blackwell, Oxford.

Locke, J 1988, *Two treatises of government* [1690], ed. P Laslett, Cambridge University Press, Cambridge.

Lloyd, G 1984, *The man of reason: 'male' and 'female' in western philosophy*, Methuen, London.

Lucy, W & Mitchell, C 1996, 'Replacing private property: the case for stewardship', *Cambridge Law Journal*, vol. 55, pp. 566–603.

MacPherson, CB 1964, *The political theory of possessive individualism: Hobbes to Locke*, Clarendon Press, Oxford.

Malafouris, L 2013, *How things shape the mind*, MIT Press, Cambridge, MA.

Malafouris, L 2015, 'Metaplasticity and the primacy of material engagement', *Time and Mind*, vol. 8, no. 4, pp. 351–371.

Merchant, C 1980, *The death of nature: women, ecology, and the scientific revolution*, Harper and Rowe, New York.

Mills, C 1997, *The racial contract*, Cornell University Press, Ithaca, NY.

Naffine, N 1998, 'The legal the legal structure of self-ownership: or the self-possessed man and the woman possessed', *Journal of Law and Society*, vol. 25, pp. 193–212.

Naffine, N 2009, *Law's meaning of life: philosophy, religion, darwin and the legal person*, Hart Publishing, Oxford.

Nedelsky, J 1990, 'Law, boundaries, and the bounded self', *Representations*, vol. 30, pp. 162–189.

O'Donnell, E 2019, *Legal rights for rivers: competition, collaboration, and water governance*, Routledge, Abingdon.

Pascoe, B 2014, *Dark Emu*, 2nd edn, Magabala Books, Broome.

Philippopoulos-Mihalopoulos, A 2011, '"... the sound of a breaking string": critical environmental law and ontological Vulnerability', *Journal of Human Rights and the Environment*, vol. 2, pp. 5–22.

Plumwood, V 1993, *Feminism and the mastery of nature*, Routledge, London.

Plumwood, V 2002, *Environmental culture: the ecological crisis of reason*, Routledge, London.

Pottage, A 2007, 'The socio-legal implications of the new biotechnologies', *Annual Review of Law and Social Science*, vol. 3, pp. 321–344.

Stolzenberg, N 2010, 'Facts on the ground', in E Penalver & G Alexander (eds), *Property and community*, Oxford University Press, Oxford.

Stone, C 1972, 'Should trees have standing?' *Southern California Law Review*, vol. 45, pp. 450–501.

Tully, J 1993, *An approach to political philosophy: Locke in contexts*, Cambridge University Press, Cambridge.

Vial, T 2016, 'Kant and Race', in T Vial (ed.), *Modern religion, modern race*, Oxford University Press, Oxford, pp. 20–53.

CHAPTER 13

Nature's Law or Law's Law? Community of Life, Legal Personhood, and Trusts

Miguel Vatter

Introduction

After years of relative disregard, the debate on legal personhood has met with renewed interest, not only in jurisprudence, but also in history and political theory.[1] There are probably many causes for this return in vogue of a topic that was last debated with some passion nearly 100 years ago.[2] One such cause is related to the discovery of biopolitics.[3] Against the backdrop of irreversible ecological catastrophe and spurred by the rise of "political ecology," the discourse of biopolitics has received a more "affirmative" meaning, in which the "community of life" of the entire earth is interpellated as a possible source of planetary law and politics.[4] Thomas Berry opposes "the inherent rights of the natural world" to the

An earlier version of this chapter was given at the "Brave New Law" Workshop, hosted by the Biopolitics of Science Research Network at the University of Sydney, August 27–28, 2018. I wish to thank Professor Brad Sherman for his very helpful commentary to my chapter.

M. Vatter (✉)
Flinders University, Adelaide, SA, Australia
e-mail: miguel.vatter@flinders.edu.au

M. de Leeuw and S. van Wichelen (eds.),
Personhood in the Age of Biolegality, Biolegalities,
https://doi.org/10.1007/978-3-030-27848-9_13

rights of corporations in industrial society (Berry, cited in Burdon 2011, p. 228).[5] The idea of "Earth jurisprudence" is slowly starting to make inroads into mainstream jurisprudence.[6] By recovering a long-standing awareness that, as one living species among others, humans have shared "biological" interests with others species, or, said otherwise, because the lives and ecosystems of all forms of life on earth are "affected" by the actions of our species (as per the Anthropocene hypothesis), the question inevitably arises as to what legal-political "standing" and legal-political "representation" nonhuman forms of life may have a right to claim.[7] The leap from acknowledging a concept of trans-species "community of life" to giving legal rights to nature's components has already been taken by the recent wave of legislative initiatives that accord natural forms of life or communities of life, like the Whanganui River in New Zealand, the legal status of citizens.[8]

But what kind of "natural" legal person is a river? Does such a river "natural" person have the same rights and duties as a human "natural" person or, instead, is it more comparable to a human-made "artificial" legal person? Simply granting legal standing to nonhuman forms of life, and thereby incorporating them into the human legal system, does not amount to recognizing them as legal persons with personal rights that human beings have to respect, protect, and implement, nor is it the same as recognizing those laws and rights that may be inherent to Nature independent of human law.[9] Thus, Berry's desideratum as to how one can bring together "human laws and Earth laws" remains an open question.

Perhaps one of the main obstacles to develop a more "ecological" conception of law is the tension that exists between the ecocentric belief in the "interconnectedness" of all life on earth—which is axiomatic in the discourse on ecological sustainability—and the prevalent tendency of human legal traditions to link rights and duties to conceptions of "personhood" that may be inapplicable to other living beings, both natural and artificial.[10] In reality, the conceptual web through which to make sense of Earth jurisprudence and Nature rights remains, to a large extent, still to be woven, despite the increasing amount of literature on an ecological approach to law. In this chapter my purpose is very preliminary and tentative. The chapter argues for introducing the perspective of contemporary biopolitics in order to think about how the requirements of Earth jurisprudence can stimulate the deconstruction of traditional jurisprudential assumptions in relation to the legal person, and to reconstruct its category in a different way.

That such an exercise may have some utility is suggested by the fact that even sophisticated jurisprudential understanding of river rights, such as found in the recent work of Erin O'Donnell, still adopt a traditional approach to the question of how legal rights are to be granted to nonhuman living beings. O'Donnell, in fact, answers the question by appealing to "the creation of 'legal fictions'," whereby a (human) legal system "creates a new, identifiable legal entity (the legal person) which includes all the legal rights (standing, contract, property) for granting the nonhuman entity its own personality."[11] But in so doing, O'Donnell simply assumes that biological life can be subsumed under the "fictional" idea of legal personhood. This is the assumption the following pages wish to question.

Nature's Law or Law's Law? Legal Personality and the Biopolitical Turn

About ten years ago now an interesting debate took place between Gunther Teubner, the sociologist of law, and Bruno Latour, the sociologist of science, over the question of how best to recognize nonhuman beings in law and politics.[12] Their approach to this question attempted to open up a third path that rejected both the traditional anthropocentric approach to law, based on the "natural" human person as ontologically prior right-holder, and the traditional ecocentric standpoint (adopted and adapted by Gaia hypothesis, "deep ecology," etc.) that is based on the recognition of Nature "herself" as a person that stands in certain "familial" relations with human and nonhuman beings.[13]

This third, sociological path to ecology consists in employing/projecting the device of "person" onto nonhuman life-forms, but depriving it of all the "humanist" baggage associated with the concepts of intentionality and responsibility (Teubner 2006, pp. 510–511). This de-humanization of the person rests on the ontological assumption of a "third" something that is neither object nor subject, something that separates and reconnects nature and culture, animality and humanity, in such a way as to undo all of these traditional dualisms. The "third" something that plays the role of "absolute" mediation in these sociologists is the idea of an autopoietic system (Luhmann, Teubner) or the idea of a network

(Latour). From this sociological perspective, anything that is nonhuman can still interact in a human-like fashion (and thus, in principle, be granted rights and duties) as long as it is functional to the communication occurring within the system or network about the environment that surrounds it.

Crucially, in both Teubner and Latour it is the device of legal personhood as a *fictional* person that plays the decisive role in extending legal and political community to nonhuman entities. Teubner emphasizes the foundational difference between "real" human and nonhuman beings and "fictional" systemic "personae": "Via structural coupling they [the fictional 'persons' of human individuals, collectives, or hybrids/MV] establish contact of communication to outside 'real' dynamic, pulsating processes, processes that occur in the environment of communication, be they flesh-and-blood-people or nonhuman processes in nature or technology. Through the masks of its 'persons,' social systems make an effective, though indirect, contact to humans and to non-humans" (Teubner 2006, p. 515). Similarly, Latour's entire "political ecology" relies on the fiction of the legal person, or what he calls the "spokespersons" that, literally, personate the nonhuman actors.[14] For Latour, the main spokesperson for nonhumans remains the "lab coat": "We shall say, then, that lab coats have invented speech prostheses that allow nonhumans to participate in the discussions of humans, when humans become perplexed about the participation of new entities in collective life" (Latour 2004, p. 67). This fictional idea of personhood allows nonhuman beings to accede to the status of "actants," defined as anything that "modify other actors through a series of trials that can be listed thanks to some experimental protocol" (Latour 2004, p. 75).

However, such actants only access human flows of communication at the price of giving up any "real" sense of life, that is, at the price of disconnecting themselves from their communities of life. Latour's ecological politics is constituted by an extreme and generalized instrumentalism: "reality grows to precisely the same extent as the work done [in experimental 'trials'/MV] to become sensitive to differences. The more instruments proliferate, *the more the arrangement is artificial*, the more capable we become of registering worlds. Artifice and reality are in the same positive column" (Latour 2004, p. 85, emphasis mine). Indeed, from the perspective of "real" living beings, the "relation of mutual experimenting" or of "mutual irritation" that characterizes systemic or network interactions between "actants" seems to indicate more a status of natural rightlessness than a case of "successful" extension of legal and political personhood beyond the human sphere, as Latour and Teubner assume. By de-humanizing personhood in order to extend its artifice to nonhuman life, Teubner and Latour have not managed to treat nonhuman life with the "inherent" or "intrinsic" value that is demanded by the ecocentric view, and that spurs the development of Earth jurisprudence and Nature rights. In other words, the attempt to give legal and political standing to nonhumans through systems and networks avoids posing the fundamental question as to the **source** of "nature" rights or "earth" law: is it nature's law or law's law?[15] By way of contrast, the ecocentric vocabulary of "intrinsic" value of nature, and the demand that human law should not "instrumentalize" nature, at least help to *pose the question* of the "sources" of law.[16]

From Teubner's and Latour's standpoint, the source of nature's law is the network or system that "projects" an artificial or fictional personality on beings that live outside of its system in order to "interact" with them. This standpoint does not ultimately differ from the traditional western idea that legal personality is posited starting from the human legal system itself, something that western legal systems have done for a long time, by giving a legal status to everything from body parts to ships. John Dewey already formulated the basic principle in his classic essay on legal personality, according to which "person" in law is just a legal construction. Basically, a legal system can make any entity, whether natural or artificial, into a legal person, if it so deems it necessary to carry out the business of law.

The main problem with this view for the development of earth jurisprudence is that the **artificial** way in which nonhuman **living** beings acquire legal standing and legal personhood within systems and networks at the same time denies them the very feature that justifies the enterprise of recognizing their legal and political status, namely, the **commonality of life** that antedates and traverses all systems and networks. This denial of the community of life in systems theory is visible in Teubner's identification of a "paradox" of animal rights. According to Teubner, animal rights are paradoxical because they "incorporate animals into human society in order to create defenses against the destructive tendencies of human society against animals" (Teubner 2006, p. 521). In other words, for Teubner it is impossible that animal rights be nature's rights: they must be (human) *law's* rights for animals rather than (human) law's rights for humans or for corporations.

Analyzed from a biopolitical perspective, the key assumption made by Teubner's formulation of the paradox of animal rights is that human society is shorthand for a *protective* or *immunitary* society. Society is constituted by the imperative to be protective to such an extent that it protects even against the damage generated by society itself, as happens when it grants animals legal standing and rights.[17] Yet, the idea of Nature rights and Earth jurisprudence calls into question Teubner's protective or immunitary assumption. The real problem is that human systems and networks are constituted as self-protecting ("autopoietic") constructions that seek to control the alterity found in their "environment." But from the perspective of the community of life, a "human" society (that is a society that represents itself as a series of "structurally coupled" systems and networks composed of human and nonhuman beings) is not an apparatus that protects from destruction but is rather an apparatus that destroys by protecting.[18]

These kinds of considerations suggest that the objectives of Earth jurisprudence are better pursued by adopting a biopolitical discourse, for it is the latter that gives access to something like a **non-immunitary source** of law. In the post-Foucaultian development of biopolitics found in thinkers like Giorgio Agamben, Antonio Negri, and Roberto Esposito, the source of law is not reduced to a system or network that projects personification in order to cope with alterity but is the reality of **shared or common life** that is presupposed by any and every immunitary prosthesis. In the terms advanced by Esposito, the source of nature's law is life as the *munus* (gift-giving obligation) that demands of all living beings that they give of themselves more than they retain for themselves.[19]

From this biopolitical perspective, the source of Earth jurisprudence cannot be law's law. Nature rights are not the rights of those "spokespersons" of nonhumans that participate in an "artificial" process of communication. Nature's law is both anterior and exterior to the legal system. Law can be considered to have a "natural" source to the extent that it emerges from the *munus* of life (or, the community of life) that is shared by any form of life, whether this form be "natural" or "artificial" (human-made). Where law's law functions by protecting the system from life as *munus* through the artifice of personality, nature's law should designate the set of duties and rights that place in community those entities that have been systematically separated into artificially established roles and given artificial scripts.

Nature Rights, Corporations, and Trusts

From its beginnings in Roman law, the idea of (human) legal personality carries within it a distinction between "natural" legal persons and "artificial" natural persons that straddles the distinction between individuals and groups. Natural legal persons can be slaves, children, women, freemen: the important thing is that they are all individuals. Artificial legal persons are usually groupings of natural persons, like churches, companies, and corporations. The radical interdependence of living beings poses a problem for an individualist approach to rights. Berry says that "rivers have river rights. Bird have bird rights. Insects have insect rights. Humans have human rights" (Berry, cited in Burdon 2011, p. 229). But, at least when rivers have been granted rights, they have been understood as a group (or collective, or assemblage) of interrelated beings: water, air, stones, forest, fauna and flora, humans, etc. If Nature rights are rights of groups, then clearly Earth jurisprudence requires questioning the traditional legal belief that groups have legal standing and legal personhood only "by fiction" and as applied to "artificial" entities like corporations. Since advocates of Earth jurisprudence, like Berry, tend to oppose corporate rights to nature rights, it is all the more important to get some clarity on the character of the group legal personality at stake here.

My hypothesis is that in order to think about the possibility of nature's law (as opposed to law's law), it is necessary to think about the possibility of a living grouping's "real" legal personality. Nature's law would then be quintessentially group law, whereas law's law (at least in modernity) is quintessentially individual or personal or subjective law. Depending on one's perspective, the jurisprudence of nature's law has either always already been devised by those peoples and cultures that preceded "modernization," or this jurisprudence has yet to be developed by future generations who shall come to realize that "we were never moderns" (Latour). However, in previous work I have suggested that some interesting notions about group law were developed during the course of the last important debate on legal personhood, and in particular in the debate between the Romanist, traditional idea of the "fiction theory of corporate personhood" and the "organic" or "group personality" theory of corporate personhood.[20] What is at stake in this and the next section is an attempt to reconstruct an "archeology" of the trust in opposition to the idea of corporation so as to recover the concept of trust as a possible figure for Earth jurisprudence.[21]

Current work on Earth jurisprudence has identified in the idea of trust one of its most promising conceptual tools. Mary Christina Wood has put forward the idea of "Nature's Trust" and Klaus Bosselmann has argued that governance of the "global commons" must be based on the idea of "trusteeship" (Wood 2013; Bosselmann 2015). For Bosselmann, "the reason why stewardship and hence trusteeship arise time and time again in religions is that common to all religions is a deity that is the true owner of the earth, for whom humans act as trustees, and who are also beneficiaries alongside future generations" (Bosselmann 2015, p. 218). Wood speaks of "a legal principle that, in some form, has flowed through countless forms of government since time immemorial" according to which "government [is designated] as trustee of crucial natural resources and obligates it to act in a fiduciary capacity to protect and restore natural commonwealth for the beneficiaries of the trust, which are present and future generations of citizens" (Wood 2017, p. 98). In this section, I hope to add a biopolitical perspective to this emerging discussion of the trust in order to question the subsumption of the trust under a sovereign relation of power that is still operative in Earth jurisprudence.

In both Wood and Bosselmann one finds a Lockean view of the fiduciary relationship, whereby the sovereign state remains both all-powerful and also a "trustee and servant" to the "people" who have reposed their "trust" in it (Locke as cited in Criddle and Fox-Decent 2009, p. 350).[22] It is this double duty, as sovereign and as trustee of the people, that Bosselmann seeks to put into use in order to remind the power of the sovereign of the "vital ecological component of that [trusteeship] role" (Bosselmann 2015, p. 179). Analogously, in Wood's terms, it is in its function of trusteeship that the sovereign can act as "a formidable counterweight to private property rights–one that situates ownership in balance with the needs of the community (both the natural and the human community)," despite the sovereign being set up essentially to protect the property rights of its members (Wood 2017, p. 110).

These interpretations of the sovereign as trust are problematic for the same reason that it is problematic when Earth jurisprudence represents the fiduciary relationship of humanity toward the community of life as being analogous to those of a parent over a child (Bosselmann 2015, p. 125, 173).[23] The image is problematic on a series of counts, but two reasons are of particular importance in what follows. The first is that basing the fiduciary relationship on the relation of parent (here, the state) to child (here, the people) subverts the republican basis of a legitimate fiduciary relationship, namely, the assumption that the people are always more powerful than the state. The second is that, in Western law, this filial-legal relationship arguably contains the roots of the destructive-because-protective function that has always been ascribed to sovereignty,[24] and that Agamben and Esposito have identified with the absolute legal power of the father over the child in Roman law that underlies all personal rights granted adults.[25] The concept of trust, if it is going to be compatible with the hypothesis of Nature rights and Earth law as discussed so far, has to be reconstructed so as to discard the filial-legal scheme in which the fiduciary relationship has been cast.

On the definition given by Innocent IV and also found in civilian lawyers like Bartolus and Baldus, a corporation or *universitas* is "a body composed of a plurality of human beings and an abstract unitary entity perceptible only by the intellect and thus distinct from its human members" (Canning 1987, p. 186). On this conception, a corporation of individuals is a group of people who have been *incorporated* into a juridical personality which is distinct from the natural persons involved in the association. Each one of these natural persons has "limited liability" with respect to this corporation. The group is perpetually represented by a *persona ficta* with its own dignity, rights and special powers of legislation. Additionally, since the personality of the group is "fictional," the *persona ficta* itself needs to be represented or embodied by one or more real, natural persons. This natural person will be "the personification of the *community in its head*" (Tierney 1983, p. 26, emphasis mine). This aspect of the corporate form of association played a crucial role in the development of the capitalist conception of the firm because it assured that the individuals who run the company, who invest in the company, and who work for the company are not personally liable to creditors of the company, or to legal prosecution in case the company incurs into violations of the applicable legal, moral or other codes in the countries in which they operate.

Two crucial consequences follow from the fiction theory of corporate personality. First, once a grouping of individuals becomes "incorporated," their lived association, their form of life in common, ceases to have a real existence in law. Common life is no longer the source of legal status, but, to the contrary, the community receives a "personality" by fiction that dispossesses the grouping qua grouping of any capacity to be self-moving and self-generating, much less any capacity of volition or intention of its own. In this sense, the incorporation of a group renders this group "stupid," and therefore considered to be ***irresponsible for its actions***. The theory of corporations thus makes groups or associations into permanent minors. This explains another important characteristic of incorporated bodies: they exist only because they are recognized by a sovereign authority. Incorporation is always a concession of the sovereign state.[26] Second, the incorporation of a living group or community of life forces the community to acquire a hierarchical organization, with a "head" (the natural person representing the legal person of the incorporated body) that commands over the "body" (the natural persons that have been incorporated into the legal person). The community of life then comes to be modeled on a body–mind dualism, where the single-mindedness of the representative's exercise control over the plural bodies of the represented. Corporations are thus a form of organization based on domination.

The canonist and civilian idea of the fictional person of groupings did not take hold everywhere in the same way. In a famous series of articles around the turn of the last century, the English historian of law Frederick W. Maitland argued that English common law developed, at the margins of Roman law, the idea of an "unincorporated body" which received the name of "trust." In Germany this kind of grouping was known with the name of a *Genossenschaft* ("fellowship") and their theorist was Otto von Gierke.[27] According to Maitland, the trust refers to a social reality, an association or community, without fictitious personality and thus without rector or guide. The existence of the grouping is directly the unlimited responsibility or liability of all members toward each other. As interpreted by Maitland, the trust or "unincorporated body" is an interesting innovation in social ontology because it appears to be the first time in modern legal history that a grouping was recognized as having a real, not a *fictional* personality.

This reconstruction of the idea of trust, in my opinion, is to be preferred to other reconstructions of the fiduciary relationship for two basic reasons. First, on this reading, the trust is a form of association that understands itself directly in opposition to sovereignty. It avoids the problem of mixing trust with sovereign that could lead to exactly the same results as mixing trust with corporation (a process that occurred in the United States during the nineteenth century and led eventually to the need for "anti-trust" legislation). Second, the principle of solidarity is preferable to the current principle of sustainability that is found in ecocentric approaches to governance because it allows one to avoid the over-arching familial metaphor of some members of the association being the "carers" for other members of the association who are deemed to be lacking in one respect or another.[28]

To make sense of the common life of groups without representatives, Gierke and the English pluralists spoke about the idea of "group personality" [*Gesamtpersönlichkeit* or *Gemeinwesen*].[29] What they seemed to be after with this term is that the grouping, as a form of life-in-common, cannot be represented by one or more natural persons, contrary to what happens with the idea of a juridical or fictional personality.[30] The personality of the group, in this sense, is articulated in the third person. The notion that a grouping can think and act without needing to be represented by a natural or artificial "person" is alien to the tradition of Roman law. But it is this very idea that allows Gierke and Maitland to affirm that a "social body without corporation" or a "community without society" should not be considered an immature, unthinking, irresponsible collection of individuals, who is "alive" only thanks to its representative "head" or "guardian." On the contrary, Maitland suggests that only un-incorporated social groupings make an authentically "common" use of reason (a truly "public" use of reason).[31] It remains an open question whether this conception of a group mind is, perhaps, better equipped to articulate the thinking processes of nonhuman forms of life, of communities of life, thereby giving the legal form of the trust a strong claim to be the more adequate form of law for nature's law.

On this reconstruction of the concept of trust, it is possible to understand trusts as a sort of political *anti-bodies*, as an *anti-corporation*. In the first place, trusts have the potential of generating an immunity against every form of sovereignty thanks to giving juridical form to the unrestricted solidarity and commonality of its members. If one adopts Esposito's claim that the Western conception of sovereignty is constituted by mechanisms of immunization of individuals from the common (subjective or personal rights being one of the central mechanisms of immunity) (Esposito 2002, 2008, 2010), then one can think of the idea of the trust as a co-immunitary and auto-immunitary resistance to such individual immunities (that is, a legal resistance to an excess of Roman law, of subjective rights, of capitalist firms, etc.). This conception of the trust could recover nowadays significance in the case of all groupings of living beings who have forms of law that emerge spontaneously from their life-in-common, and yet who are not recognized as being "sovereign" by the organization of the sovereign state, in part because their law does not fit with the form of positive law issued by a sovereign authority.

Trusts and the Separation of Use from Dominion

The second sense in which trusts are anti-corporations can be evinced from the role they played in the development of western conceptions of private property. Ecocentric jurisprudence, particularly the work of Prue Taylor, has emphasized the need for a "new vision of property" that would be more appropriate for the "biotic community."[32] Based on the "land ethic" proposed by Aldo Leopold, the emphasis in this new vision of property would be on the obligations toward the land and its ecosystems that are comported by the right of private property. On this view, "property rights per se are not the issue–rather the issue is what we understand property rights to entitle us to do and what obligations they impose" (Taylor and Grinlinton 2011, p. 5). Taylor introduces here a distinction between anthropocentric interests in private property based on growth and ecocentric interests of the environment based on sustainability and the "ecological limits of the Earth" (Taylor and Grinlinton 2011, p. 13), so that rights to private property are intimately correlated with "delegated responsibility for land" (Taylor and Grinlinton 2011, p. 16). Ultimately Taylor refers to a combination of "relational" and "obligational" theories of private property. However, the connection between forms of life and modes of property is not thermalized as such, and neither is the connection between trusts and private property.[33]

To recover these connections, it is necessary to pursue the archeology of the trust and its relation to the history of private property in the West. Historically, the concept of the corporation or "incorporated body" originates in the context of a conflict between the Papacy and the Franciscan movement over the legitimacy of private property. The Franciscan movement implicitly called on the Church, and on the Pope as its representative, to return to the original form-of-life of Jesus and the Apostles, which they thought entailed giving up "dominion" over all worldly things. Members of the Franciscan Order therefore made a vow of "highest" poverty. They gave up not only the **right to own** something, but even the **right to use** anything, and only kept the "natural right" to the **actual use** of things for the sake of preserving their "bare" life. They believed that this "natural right" was shared by all living creatures in their community of life. One could perhaps say that the ideal of "highest poverty" is the first appearance in modern western history of something like "nature's law."[34]

The history of the Papal response to the Franciscan challenge is undoubtedly the most significant legacy bequeathed to modernity by medieval Scholastic legal and political thought because it has been seen by many as the birth of modern conception of private property and individual or personal rights.[35] In Roman civil law, the use of anything required dominion over it. All things, in so far as they are not persons, must belong to some person: to be used, any object must first come under the dominion of a person. In other words, to be a person (in law) means to become an owner.[36] The Papal lawyers exploited the paradox that Franciscans used up a lot of things, as if they owned them, even though nominally they were poor and owned nothing. The principle that was developed by these lawyers in their combat against the Franciscan ideal of "highest" poverty is none other than the idea that the use of something requires not only dominion over it but also its consumption. The Pope's lawyers responded to the Franciscans that, essentially, to make use of something is to consume it.

For Giorgio Agamben, the failure of the Franciscan movement to defend their conception of "highest poverty" in juridical terms sealed the internal connection between use, ownership, and consumption that characterizes capitalist property relations and the primacy of the right to private property as basis of individual freedom. Contemporary philosophers like Negri and Agamben have turned to the Franciscan idea of "highest poverty" in order to advocate for the anti-capitalist idea that there are things which *can only be used in common* and that *cannot be owned by any person*. This idea, which Negri and Hardt refer to as the "common," should be of central importance to the development of Earth jurisprudence, in particular because it is internally linked with the on-going shift of the capitalist industrial economy toward a new *bio-economy*. For Negri, a bio-economy entails the transformation of work from being the activity of generating *products (commodities to be consumed)* to be the activity of providing *services (forms of life to be sustained)*. A "service," on this definition, is nothing other than what allows someone to *make use of something without owning it*.[37] This biopolitical recovery of Franciscan "highest poverty" is also directly pertinent to the situation of the "global commons" in the Anthropocene, for clean drinkable water and breathable air, biodiversity, ocean temperatures, parts of space and the atmosphere, some cultural goods (like temples, art, palaces, urban or natural landscapes), and some natural living beings (like rivers, barrier reefs) are all things that cannot be owned by anyone because they are intended for the universal use of all living species.

The conception of trusts discussed above connects with these bio-economical developments by proposing **an entirely different connection between use and ownership** than the one established in Roman law and then adopted and adapted in modernity. Trusts lend themselves to a new use in the context of the Anthropocene bio-economy. In his celebrated essay "Trust and Corporation," Maitland argues that the concept of trust is born in the context of a struggle about property, much like its antonym, the idea of a corporation. But whereas the corporation is a legal instrument that favors the class of the owners (those who have *dominium*), Maitland suggests that the trust was a legal instrument that favored those who had little or nothing other than their bare lives. At the time of the Papal introduction of the idea of corporation as fictional personality, the problem facing the peasantry was that anything someone "used" (for example, a house, a field, cattle, etc.) could not be given as inheritance to someone else in their family by means of a will (testament). So, the question arose: How can one pass on the fruits of social labor down to the next generation without being their legal proprietor?

According to Maitland, the trust was invented to address this problem. The solution lies in creating a "trust" which contains use-values, and in designating a group of friends or associates, the *trustees*, who shall administer this property *for a third person*. Trustees are to look after this property *for the beneficiary* of the trust, namely, for the person who is designated to inherit the goods, and this person can even be oneself considered as a third person.[38] These trustees can *own* the trust *without thereby making use of it for their own profit* (for example, trustees cannot "sell" the property to someone else who is not the original beneficiary of the trust). In order to protect the beneficiary, the buyer of any property was supposed to ask for the title deeds, to ensure that he would not buy a property sold by a trustee as if they were the beneficiary. In this way the trust can be transmitted from generation to generation, incremented in time (if the trustees are not irresponsible), saved from creditors (if the trustee owes money to someone, they cannot take it from the trust), so that property may be *used by a third person* (the beneficiary) *without being consumed by its owner* (the trustee).

Traditionally, the trust is thought to give "rise to two types of ownership of property simultaneously in existence... called legal title and equitable title" (Bosselmann 2015, p. 138). But this way of interpreting the trust does not quite do justice to its revolutionary potential. For the legal structure of the trust entails a high degree of impersonality: what is "mine" becomes "yours" (for example, it becomes the property of the trustees, who have the legal title of dominion over the goods), but "you" (the trustees) "own" the goods only in accordance to common law, and must hold these goods "to his use [*a son oes*]" of a third person, namely, the beneficiary of the trust, and this in accordance to *equity* law.[39] In the trust, the third person becomes the subject of *justice as equity* (whereas in common law, justice is purely *distributive and commutative*, ultimately a matter of guilt and punishment, of debtors and creditors, of first and second persons). In other words, and entirely contrary to the Roman connection between use, dominion, and consumption, the trust allows some third person to use what is owned by me; what is owned by me cannot be consumed by me because it places me under the obligation to sustain it. Lastly, I do not have dominion over what is owned by me: my actions are to be judged according to considerations of equity, rather than by considerations of distributive or commutative justice that apply to private property and their owners. The trust is, therefore, more than a second type of property: it is rather a legal way to radically distinguish use from dominion and ownership that opens up the domain of use to an indefinite "third person" defined as any member of a community of life.

Returning, in conclusion, to the first part of the chapter, my hypothesis is that this idea of the trust could be employed to articulate the two contrasting approaches to the source of law that was discussed above: what I called "nature's law" and "law's law." These two sources would give rise to the dualism of equity and common law, respectively. What is to be used without being owned by the community of life on earth becomes, in such trusts, the property of trustees held for the present and future use of "a third person," which is a group personality of beings living on earth. Conflicts between subjective or personal rights of human beings and rights of nature would have to be resolved by specially instituted courts of equity based on a conception of human dignity and human rights that have been enlarged to include all living species, be these natural or artificial (for example, future generations of intelligent and self-learning machines, creations of synthetic biological hybrids, etc.).

Notes

1. For the debates in legal theory, see Naffine (2009), Grear (2013), Parsley and Mussawir (2017), Van Beers (2017), among others.
2. For a recapitulation and attempt to bring closure to that earlier debate, see Dewey (1926).
3. For a collection of recent approaches to biopower and biopolitics, see Vatter and Lemm (2014).
4. For two very different approaches, see Berry (1999) and Latour (2004).
5. The opposition between neoliberalism and an alternative "earth governance" based on the rights of Nature is also one of the leitmotifs in Bosselmann (2015).
6. For "earth jurisprudence" see the materials collected in https://www.earthlaws.org.au/what-is-earth-jurisprudence/rights-of-nature/.
7. In accordance to the fundamental Roman legal principle *quod omnes tangit*. This Roman legal principle seems to underlie Latour's recommendation that we shift from thinking about non-humans not as "objects" or as "matters of fact" but as "matters for concern" (Latour 2004, p. 22, 66). Latour does not avow this Roman legal inheritance.
8. See Charpleix (2018) and literature cited therein. For questions of the effectiveness of nature rights on environmental protection, see Whittemore (2011), Good (2013) and O'Donnell and Talbot-Jones (2018).
9. For the debate on "legal standing" of nonhumans, see also Grear (2012).
10. For some interesting considerations, see now Hayles (2017). On the deep-seated relation of law to personality, see in general Esposito (2015).
11. Talbot-Jones (2018) and now O'Donnell (2018, Chapter 7, passim).
12. See Teubner (2006). On Latour's ecological politics, see now McGee (2014).
13. Perhaps the iconic symbol of this in recent earth jurisprudence is the conception of Mother Nature or Pachamama in the Bolivian push for a "Universal Declaration of the Mother Earth Rights," but equivalents are to be found in most First Peoples cosmologies, see Cullinan (2003). For a more in-depth look at the Maori case, see Thomas (2011).
14. "I do not claim that things speak 'on their own' since no beings, not even humans, speak on their own, but always through something or someone else" (Latour 2004, p. 68).
15. I am suggesting that Latour's provocative call to "let go of Nature" so as to attain a truer "politics" of nature, or political ecology, confines him to a technocratic, prosthetic conception of law that undoes any "ecological" claims he may put forward for his politics of hybrid (human and non-human) collectives. See his critique of "Nature" in Latour (2004, pp. 28–29).

16. On the other hand, the ecocentric view often mistakenly articulates this insight by adopting, in turn, a very traditional, humanist belief that seeks to deduce law from (human) morality.
17. For similar approach, see Wolfe (2012).
18. This is the basic critique of systems theory elaborated by Esposito in his works Esposito (2010, 2011).
19. See Esposito (2008). For a general discussion of these developments in the theory of biopolitics, see Vatter (2017).
20. I refer to Vatter (2013, 2016). I draw on these works for what follows.
21. For this concept of archeology, see Agamben (2008). On the later development and use of trust theory in European colonialism and imperialism, from Vitoria to Burke to the Mandate System of the League of Nations, see the remarks of (Criddle and Fox-Decent 2009) who argue that "the wrongfulness of colonialism lies not in the trust-like structure of colonial rule per se, but in colonialism itself." It will be clear that the concept of trust I develop is useless for settler colonial purposes.
22. See the discussion of trusteeship and sovereignty in Bosselmann (2015, pp. 169–177).
23. Bosselmann trades on the root meaning of "eco", from the Greek *oikos* or household.
24. It is sufficient to remind oneself of Carl Schmitt's formulation of the key principle of sovereignty: *protego ergo obligo*.
25. In Roman law, "no human being was a person by nature – not as such. Certainly not the slave, but not the free man either: before he became a *pater*, that is, a legal subject, he still had to pass through the status of *filius in potestate*, which reflected the fact that, in the changeable dispositif of the person, since human beings arrived into life from the world of things, they could always be thrust back into it" (Esposito 2012, p. 79).
26. See the detailed discussion of this point found in Runciman (2005).
27. I refer to the new edition of these essays in (Maitland 2003). See also (Gierke 2002). This volume is a translation of selections from *Das deutsche Genossenschaftsrecht*, mainly coming from volume 1, *Rechtsgeschichte der deutschen Genossenschaft*.
28. For a discussion of the principle of sustainability, see Bosselmann (2017).
29. In the "Editor's Introduction" Black gives the usual definition of "group personality: "Groups really have a personality, a mind and a will, and the state and the law ought to recognize this" (p. xvi). As I shall argue, the view that a group has a "personality" analogous to that of an individual is misguided, and is not what Gierke meant.
30. On the logic of "third person" see Esposito (2012). In Maitland, the point that a trust is always made so as to give "rights to a third person" is found in "The Unincorporate Body" (Maitland 2003, pp. 54–55).

31. Pettit has argued for the need to recognize that groups have minds of their own, over and above the minds of their members, and hence can also be held "responsible" for their actions or omissions. However, Pettit's reasoning seems to remain caught within the assumption that groups can only be constituted via incorporation (see List and Pettit 2006; Pettit 2003). For fascinating discussions of group personality of trees and non-humans from an anthropological perspective, see (Kohn 2013) and (Descola 2014).
32. See the general presentation found in Taylor and Grinlinton (2011).
33. These connections remain underspecified also in Bosselmann (2011) who connects a traditional, corporation based conception of private property to the demands of sustainability, but does not address the trust based conception of property.
34. On these concepts, see now Agamben (2013).
35. The story is recounted in Tierney (1997).
36. On the distinction between medieval and modern subjective rights, see Coleman (2005).
37. For a discussion of the new economy of the "commons," but without any biopolitical or ecological articulations of these insights and practices, see now Helfrich (2014).
38. The classic definition is that "the common law's trust is a strange hybrid of property and obligation; indeed, it is an obligational relationship with respect to property that has been allowed to affect third parties and has, therefore, in a most un-civilian way given property-like characteristics to the obligational claims of the beneficiaries" (Smith 2012, pp. 258–259). Bosselmann's describes it as follows: "The legal concept of the trust in the common law is predicated on a relationship between the settlor (donor) and the trustee with regard to the administration by the latter of the property or assets for the benefit of other persons, the beneficiaries or in respect of a particular charitable purpose" (Bosselmann 2015, p. 138).
39. See the discussion of "use" in "Trust and Corporation" (Maitland 2003, pp. 84–80).

Bibliography

Agamben, G 2008, *Signatura Rerum. Sul metodo*, Bollati Boringhieri, Torino.

Agamben, G 2013, *The highest poverty: Monastic rules and form-of-life*, Stanford University Press, Stanford.

Berry, T 1999, *The great work: our way into the future*, Bell Tower, New York.

Bosselmann, K 2011, 'Property rights and sustainability: can they be reconciled?' in D Grinlinton & P Taylor (eds), *Property rights and sustainability: the evolution of property rights to meet ecological challenges*, Brill, Leiden, pp. 23–42.

Bosselmann, K 2015, *Earth governance: trusteeship of the global commons*, Edward Elgar, Cheltenham, UK.

Bosselmann, K 2017, *The principle of sustainability: transforming law and governance*, Routledge, London.

Burdon, P (eds) 2011, *Exploring wild law: the philosophy of earth jurisprudence*, Wakefield Press, Adelaide.

Canning, J 1987, *The political thought of Baldus de Ubaldis*, Cambridge University Press, Cambridge.

Charpleix, L 2018, 'The Whanganui River as Te Awa Tupua: place-based law in a legally pluralistic society', *The Geographic Journal*, vol. 184, pp. 19–30.

Coleman, J 2005, 'Pre-modern property and self-ownership before and after Locke: or, when did common decency become a private rather than a public virtue?' *European Journal of Political Theory*, vol. 4, no. 2, pp. 125–145.

Criddle, EJ & Fox-Decent, E 2009, 'A fiduciary theory of Jus Cogens', *Yale Journal of International Law*, vol. 34, pp. 331–387.

Cullinan, C 2003, *Wild law: a manifesto for earth justice*, Green Books, Devon.

Descola, P 2014, *Beyond nature and culture*, University of Chicago, Chicago.

Dewey, J 1926, 'The historic background of corporate legal personality', *Yale Law Journal*, vol. 35, no. 6, pp. 655–673.

Esposito, R 2002, *Immunitas*, Einaudi, Turin.

Esposito, R 2008, *Bios: biopolitics and philosophy*, University of Minnesota Press, Minneapolis.

Esposito, R 2010, *Communitas: the origin and destiny of community*, trans. T Campbell, Stanford University Press, Stanford.

Esposito, R 2011. *Immunitas, the protection and negation of life*, Polity Press, Cambridge.

Esposito, R 2012, *The third person*, Polity Press, London.

Esposito, R 2015, *Persons and things: from the body's point of view*, Polity Press, London.

Gierke, O von 2002, *Community in historical perspective*, ed. A Black, Cambridge University Press, Cambridge.

Good, M 2013, 'The river as a legal person: evaluating nature rights-based approaches to environmental protection in Australia', *National Environmental Law Review*, vol. 1, pp. 34–42.

Grear, A (ed.) 2012, *Should trees have standing? 40 years on*, Edward Elgar, Cheltenham, UK.

Grear, A 2013, 'Law's entities: complexity, plasticity, justice', *Jurisprudence*, vol. 4, no. 1, pp. 76–101.

Hardt, M 2009, 'Politics of the common', Contribution to the Reimagining Society Project hosted by Z Communications, Boston, 6.

Hardt, M & Antonio, N 2009, *Commonwealth*, Harvard University Press, Cambridge.

Hayles, NK 2017, *Unthought, the power of the cognitive nonconscious*, University of Chicago Press, Chicago.

Helfrich, S (ed.) 2014, *Commons, Für eine neue Politik jenseits von Markt und Staat*, Heinrich-Böll Stiftung, Bonn.

Kohn, E 2013, *How forests think: toward an anthropology beyond the human*, University of California Press, Berkeley.

Latour, B 2004, *Politics of nature: how to bring the sciences into democracy*, Harvard University Press, Cambridge.

List, C & Pettit, P 2006, 'Group agency and supervenience', *Southern Journal of Philosophy*, vol. 44, pp. 85–105.

Maitland, FW 2003, *State, trust and corporation*, ed. D Runciman & R Magnus, Cambridge University Press, Cambridge.

McGee, K 2014, *Bruno Latour, the normativity of networks, Nomikoi, critical legal thinkers*, Routledge, London.

Naffine, N 2009, *Law's meaning of life, philosophy, religion, darwin and the legal person*, Hart Publishing, Oxford.

O'Donnell, E 2018, *Legal rights for rivers: competition, collaboration, and water governance*, Routledge, London.

O'Donnell, E & Talbot-Jones, J 2018, 'Creating legal rights for rivers: lessons from Australia, New Zealand, and India', *Ecology and Society*, vol. 23, no. 1, p. 7.

Parsley, C & Mussawir, E 2017, 'The law of persons today: at the margins of jurisprudence', *Law and Humanities*, vol. 11, no. 4, pp. 44–63.

Pettit, P 2003, 'Groups with minds of their own', in F Schmitt (eds), *Socializing metaphysics*, Rowman & Littlefield, New York, pp. 167–193.

Runciman, D 2005, *Pluralism and the personality of the state*, Cambridge University Press, Cambridge.

Smith, L 2012, 'The re-imagined trust', in L Smith (ed.), *Re-imagining the trust: trusts in civil law*, Cambridge University Press, Cambridge, pp. 258–273.

Taylor, P & Grinlinton, D 2011, 'Property rights and sustainability: toward a new vision of property', in D Grinlinton (ed.), *Property rights and sustainability: the evolution of property rights and sustainability*, Brill, Leiden, pp. 1–20.

Teubner, G 2006, 'Rights of non-humans? electronic agents and animals as new actors in politics and law', *Journal of Law and Society*, vol. 33, no. 4, pp. 497–521.

Thomas, N 2011, 'Maori concepts of Rangatiratanga, Kaitiakitanga, the environment, and property rights', in D Grinlinton & P Taylor (eds) *Property rights and sustainability: the evolution of property rights to meet ecological challenges*, Brill, Leiden, pp. 219–248.

Tierney, B 1983, *Religion, law, and the growth of constitutional thought 1150–1650*, Cambridge University Press, Cambridge.

Tierney, B 1997, *The idea of natural rights*, William B. Eerdmans Publishing, Grand Rapids, MI.

Van Beers, B 2017, 'The changing nature of law's natural person: the impact of emerging technologies on the legal concept of the person', *German Law Journal*, vol. 18, no. 3, pp. 560–593.

Vatter, M 2013, 'Il-limitato e s-corporato. Dalla corporazione al comune passando per il trust', *Filosofia Politica*, vol. 3, pp. 3–26.

Vatter, M 2016, 'Law and life beyond incorporation, Agamben, highest poverty and the papal legal revolution', in D McLoughlin (ed.), *Agamben and radical politics*, Edinburgh University Press, Edinburgh, pp. 234–262.

Vatter, M 2017, 'Community, life, and subjectivity in Italian biopolitics', in S Prozorov & S Rentea (eds), *The Routledge handbook of biopolitics*, Routledge, London and New York, pp. 123–140.

Vatter, M & Lemm, V (eds) 2014, *The government of life: Foucault, biopolitics, and neoliberalism*, Fordham University Press, New York.

Whittemore, ME 2011, 'The problem of enforcing nature's rights under Ecuador's constitution: why the 2008 environmental amendments have no bite', *Pacific Rim Law & Policy Journal*, vol. 20, no. 3, pp. 659–691.

Wolfe, C 2012, *Before the law, humans and other animals in a biopolitical frame*, University of Chicago Press, Chicago.

Wood, C 2013, *Nature's trust: environmental law for a new ecological age*, Cambridge University Press, Cambridge.

Wood, C 2017, 'The nature's trust paradigm for a sustaining economy', in MK Scanlan (ed.), *Law and policy for a new economy*, Edward Elgar, Cheltenham, UK, pp. 97–115.

CHAPTER 14

Afterword: After the Great Undoing

David Delaney

This is an inspiring, provocative collection of papers that each address a question of vital urgency: who, or what, are we becoming? The chapters gathered together in *Personhood in the Age of Biolegality: Brave New Law* investigate many of the key conundrums and lines of fracture that attend this, in many ways unprecedented, moment in the history of modern personhood. While the chapters touch on a wide range of perplexing topics from data doppelgangers, genetic privacy and the rights of pre-conceived children, to contending theories of pain and of proof, taken as a whole the volume makes a strong case that in our time the concept of personhood has become profoundly destabilized. We could say "undone". Indeed, the degree of incoherence may be such as to more accurately support a diagnosis of "deranged". Ghosh (2016) refers to the present post-Holocene moment of planetary history, more generally referred to as the Anthropocene, as the Great Derangement. Arguably, the initiation of this epoch was co-incident with and causally related to the invention of the modern person.

D. Delaney (✉)
Department of Law, Jurisprudence and Social Thought,
Amherst College, Amherst, MA, USA
e-mail: dpdelaney@amherst.edu

M. de Leeuw and S. van Wichelen (eds.),
Personhood in the Age of Biolegality, Biolegalities,
https://doi.org/10.1007/978-3-030-27848-9_14

In our world, "person" is not a minor concept. For people like us it is as good a candidate as any for being regarded as fundamental. Therefore, the entities to whom it is applied (*or* from whom/which it is withheld or withdrawn) might likewise be deranged. And why not? If we assume, first, that the inherited notion of the person (that is, a version of the Cartesian, disembodied, autonomous, minded individual) is now under siege from an array of opposing positions; second, that a successor person (or "entity") has yet to achieve hegemonic status; and third, that in order to competently function in a social world—in order to learn how to *be* a person—some relatively stable, hegemonic model needs to be internalized, then, as the inherited version disintegrates under the force of an ever growing list of contingent possible replacements, of course some of us will *feel* deranged! A central question posed by the volume is: how might what we call law respond to this disintegration? Or, put another way: how does metaphysical dissensus translate into politics and how does the politics of personhood revision get re-translated into the idioms of rules and rights? How can law even maintain its always deferred dream of coherence when one of its foundational underpinnings is, like a broken mirror, crazed?

Any discussion of contemporary re-workings of personhood must recognize that "person" is intrinsically a powerful concept. As implied in many of the chapters, modern personhood has always been characterized by multiplicity, plasticity, and contestability. Gaps and overlaps of experienced personhood, legal personhood, full or partial personhood, natural and artificial flavors of personhood are refracted through the politics of race, gender, divergent corporealities, citizenship, labor, science, commerce, and more. Power relations are implicated in the determination of those entities to whom the status of person is granted, in full or part, or withheld, and in the determination of the criteria of membership. Such determinations position these entities within relational grids that condition the distributions of responsibilities, rights, and no-rights, that is, power-to or power-over. The origin of authority to so grant, withhold, determine and position is also a matter of power: who gets to say who is or is not a person and why? So much of the history of modern political life directly implicates struggles over inclusion and exclusion from the charmed circle of personhood and its statist inflection, citizenship. The trajectory of many of these struggles fits comfortably, and comfortingly, within the narrative frame of liberal progress. This is the story of achieved social recognition of the transformation of objects into subjects and of subjects as free agents who, through the instrumentality of rights, may participate in the satisfaction of desires in relation to other persons and, crucially, in relation to those entities who remain under the sign of objects or things. The greater the zone of freedom for each designated person, the greater the zone of freedom for the species as a whole in its eternal struggle against the constraints and compulsions of "nature"—the ultimate non-person. This simply *is* the liberal progress story. And one of its mighty engines of transformation is "the person" as so imagined.

A core feature of modern personhood, then, is its estrangement from, and opposition to the other-than-human world. An enormous amount of intellectual labor has been devoted to critiquing this view. For example consider Putnam's maxim "the mind and the world jointly make up the mind and the world" (1981, p. xi). This slogan is amenable to a number of interpretations depending on what we take "mind", "world", and "make up" to mean. Many of the chapters here can be taken as illustrations of how this works. One productive reading relies on the somewhat suspect but still serviceable concept of "culture" as implicating both "world" and "mind". Different cultural formations might be said to have or construct distinctive conceptions of minds, mindedness, or analogous understandings of interiority and, in turn, those entities so constructed—call them persons—reproduce aspects of the culture by believing themselves to have these kinds of minds or be these kinds of persons. Taking Gergen's notion of "culturally constructed conceptions of mind" (2010, p. 768) seriously we may loosely speak of Olmec, Yorta Yorta, Babylonian or late-modern (conceptions of) mind as "kinds of minds". People for whom reincarnation or the divine spark are uncontestable truths might be expected to have different conceptions of mind, and hence personhood, than those who don't.

In our cultural-historical moment, the figures of the self-interested, rational, utility maximizer or the liberal subject presuppose a kind of mind that underpins being a kind of person. As Taylor (1989) has shown, this kind of person became hegemonic in the culture that invented it. It was then disseminated throughout the range of humanity by the forces that destroyed many of the other cultural ways of being human. "Be this kind of person, or else!" This is the normative model of personhood to which responsibilities, rights and no-rights are attributed. In our world, though, one (many) may simultaneously feel themselves to be both some approximation of this self and to be somewhat, or significantly, other than this; to fall somewhat, or significantly, short of the mark. One may be held to account for non-conformance. At the most general level this captures one interpretation of the world and the mind making each other up. The chapters provide evidence that it is *this* kind of person that is being undone both culturally and experientially.

At a more detailed level of analysis Hacking's historical-ontological account of "making up kinds of people" and the reciprocal "looping effect", whereby people come to see themselves as new kinds of people, may be more useful for framing the events discussed in many of the chapters (2002). Hacking discusses the invention-cum-discovery of "kinds of people" such as perverts, homosexuals, and "the obese". For example, with respect to homosexuals he argues "'[t]he homosexual' is a kind of person that exists only in a particular historical and social setting, for example now, but not in ancient Athens. The homosexual 'as a kind of person' did not exist then, although there was plenty of same-sex acts with complex codes about which acts were rights and which were wrong" (2007, 295). And the same cultural-psychological "people-making" processes are in play with respect to, say, the spectral person, the plastic person as well as adult children of alcoholics, hipsters, neuro-typical gender-queer eco-anarchists, experts, and the seemingly infinite "kinds of people" that proliferate in—and now populate!—our late-modern world. *They are us.* This proliferation may be the most obvious evidence of the Great Undoing of personhood in our time. Something like this process is suggested by Van Beer's discussion of Adler's notion that people have come to see themselves as shamefully deficient compared to their machines, by Rabet's critique of neoliberalism's (ir)-responsibilization projects, in Addison's study of coming to see aging (that is, living) as a disease. It is, I think, implicit in van Wichelen and de Leeuw's analysis of the risks posed by the de-anthropomorphization of personhood entailed by Grear's proposed "entityhood". What happens to us when we come to accept these alternative framings as true?

As slogans such as "you are your genes", "you are your synapses", "you are your microbiome" and so on multiply, circulate and take root in the very minds they purport to be describing, *how* it is to be a person changes and Hacking's looping effect kicks in. With respect to figures such as the "cerebral subject" or "the synaptic self", Slaby and Gallagher assert that "...people ... begin to see themselves as the very neuronal machines, the very molecular automatons championed by the neuro-cognitive sciences" (2014, p. 46). They warn that "...regardless of the truth of neuro-philosophical claims, our social world is already on the way of becoming a neuronal world" (p. 44). At the same time, while some (numbers of) people may take comfort in the expert discourses of corporealization, others may feel that their first-person accounts of being a person have been illegitimately invalidated by authorities in deference to expert scientistic knowledge claims. This is evident in Barker's research on contested theories of pain, Trundle's chapter about rival theories of suffering and Ihar's on the politics of proof of age in asylum cases. Each of these implicate the authoritative erasure of first-person accounts of personhood. They also have significant consequences for the life trajectories of actual human beings under erasure.

In our moment of undoing, the cultural revisions of mind, self, personhood cannot be disentangled from companion conceptions of body and of world. Again, what is coming undone is the inherited, modern person predicated on an image of mind as immaterial, will as uncaused, and the uncaused as signifying freedom. This (kind of) person lives, works, loves, and suffers in a world that is politically and economically *predicated on* and *organized around* this modality of personhood and the valorization of desire-satisfaction and self-expression associated with it. The formal distribution of rights and, hence, power, is likewise inseparable from this metaphysical view. As this person dissolves and fragments, its formerly hegemonic, common-sense status is coming to be seen as mere ideological vestige. But again, without a clear consensus replacement the undoing of modern personhood may be regarded as entailing nothing less than a mental breakdown at the level of the species.

The inherited person is most clearly challenged by highly capitalized, scientistic revisionist projects such as those associated with AI (Parry's virtual persons), genetics, psychopharmacology, and neuroscience. These tend to be strongly materialist contenders and are commonly criticized by opponents as being reductionist and/or determinist. Perhaps the most extreme position is that which calls itself eliminative materialism which seeks to eliminate (belief in) "mind" as such (Rachlin 2014). These visions are predicated on the denial of mind-world separation. It's matter and only matter wherever you look. But while presented in the idioms of objectivist facticity, these *too*, of course, are cultural (political) productions. Anxieties about the looping effect of these revised persons have already been mentioned. As Gergen asks: "if such explanations are embraced by society – treated as true beyond culture and history – what are the implications for social practices and policy [read 'law' or 'rights']?" For him, the risk is that "...corticalization of human behavior robs human traditions of their significance" (2010, p. 798).

But resistance to the enthronement of "the synaptic self" or the "cerebral subject" comes not only from those who would retain and reinforce the disintegrating inherited person of late-modernity. Other visions of mindedness, and hence personhood, also vie for our allegiance. Other mind-maps are being drawn. On the one hand, this is simply more evidence of the proliferation and undoing of personhood in our time—more dispatches from the ongoing brain wars of late-modernity. These too are cultural "making-up-people" projects. But on the other hand, *something* has to replace the disappearance of the modern person. As mentioned, much effort and investment is being devoted to the project of establishing "the cerebral self" and brainhood as the reigning Brave New Person of the future. However, it *could* come to pass that as humanity (and other life-forms) move deeper into the Anthropocene one of these, now marginal, perhaps insurgent, mind-maps *does* achieve cultural coherence and stability. It may come to pass that one of them pre-figures the kind of persons our descendants will feel themselves to be. It may come to pass that one of them accomplishes a degree of hegemony such that it is assumed to be true by (whatever becomes of) law. If the looping effect is an enduring phenomenon, it's conceivable that future persons might decide to "make up" who they will be in accordance with one of these alternatives.

The range of insurgent mind-maps can be distinguished by how far "minds can extend beyond the boundaries of skin-and-skull" (Gallotti and Heubner 2017, p. 251), and by the extent to which "mind" remains located in "the individual". For adherents of "the embodied mind" cognition and other mental states are not limited to brain states. Rather, "...the presence of human minds depends directly upon the possession of a human like body" (Clark 2008, p. 43). Those who champion "the extended mind" approach hold that mental activity implicates dynamic participation of artifacts. In the relevant literature reference is made to notebooks, maps and the like but clearly this is familiar in the ways in which many people now "think with" their handheld devices and other AI interfaces. These are no longer mere "things" offer affordances that enable and constrain what mind can do when it thinks, remembers, intends or perceives. Others push the boundaries of mind further out into the world, perhaps dissolving the mind/world distinction that

Putnam merely made interactive. In some formulations the word "environment" stands in for "world", as when Kono writes, "...mind extends to the whole system of brain-body-environment and the self is fundamentally open to the surroundings" (p. 55). But, in shifting the referent of "world" from environment to ecology, Bragg (1996) advocates the construction of an "ecological self" that aligns with the program of deep ecology. These creative interventions should also be regarded as the activities of (culturally constituted) minds engaged in the work of re-making worlds.

Other critics insist that we commonly participate in shared intentional states with particular others through mutual capacities for attunement and alignment. For Gallotti and Huebner, acceptance of the notion of "collective minds" means that we "...cease to act as isolated individuals and begin to act together as a unified locus of thought and agency (us)" (2017, p. 256). As with Davies' engagement with the notion of distributed minds in the present volume, others map mind as networked across larger collectives., for example, Gallotti and Huebner insist that "...[W]e are deeply social organisms who cannot think, act or understand except through our socially situated, socially scaffolded and socially organized way of being in the world with others" (p. 264). They even assert that, "Such ... a system of cognitive processes could be called an agent in its own right" (p. 259). Extending the location of mind even further from individual skulls Gallagher and Crisafi's notion of "mental institutions" (2009) holds that social institutions such as law, museums—we could extend the list indefinitely to include schools, prisons, churches, corporations, workplaces—are as intimately involved in cognitive processes as are neurons. Cognitive processes, they argue "...are ongoing and outside of any particular person's head" (p. 49).

If this is the case, then it seems to entail a radically different conception of "the person" than one in which mental states can *only* take place "in the head" of an individual and *that* is what makes her a person. Davies is surely correct in her suggestion that new ways of thinking about mind and personhood necessitate the invention of "alternative legal mythologies". But would "... a view of law that is ontologically extended and distributed, like human beings" or "... a view of law interconnected with ecosocial and biophysical relations..." be recognizable, to those who live in a world founded on such views, as what we call "law?" Or might the denizens of this world be perplexed by the questions like: "what was law?" And "what were persons?" It is hard for *this* modern person (me) to discern these Brave New Persons as other than transitory locations or moments in a complex "mental" circuitry. Indeed, Di Paoli goes so far as to insist that "cognition has no location" (2009, p. 19). At this point the reciprocal, mutually constitutive dynamics of "culture" and "mind" that might inform Putnam's maxim vanishes. Culture and mind seem to be, if not synonymous, then perhaps bearing a relation similar to that of ocean and water. My very limited aim here has been to gesture toward a set of contemporary, contending mind-maps that challenge the metaphysical cartographies of *both* the inherited atomistic, immaterial minded "person" and its seemingly ascendant reductionist replacements. I should stress that I am no more competent to assess the merits of these positions than I am to assess neurocentric claims on their own terms. The point is simply to suggest the broadening cultural horizons of thinkablity.

These thoughts were inspired by my reading of *Personhood in the Age of Biolegality: Brave New Law* and the various ways that the authors have wrestled with the evidence of the undoing of the modern person. At the end of the day I ask myself, how should I know? There may be mind-maps that I find myself preferring to others insofar as I am able to imagine what, *for me*, are more or less desirable social consequences of our "making up kinds of people" who might be less destructive of "the world" and of each other than the modern self-interested, rational, utility maximizer has proven to be. If the looping effect holds, I can imagine the arrival of a normative, hegemonic (and legally enforceable?) "relational-ecological self" that might be more appropriate for living on a post-Holocene planet. But if "the world and the mind [*do*] jointly make up the world and the mind", if the world *has* become deranged; and if the Anthropocenic derangement was inaugurated by the coronation of the modern individuated person accessorized, as it is, with insatiable desires and freedoms dressed up in rights, then how could the beings to whom this personhood and this broken world were bequeathed be other than undone themselves?

References

Bragg, E 1996, 'Towards ecological self: deep ecology meets constructionist self-theory', *Journal of Environmental Psychology*, vol. 16, pp. 93–108.

Clark, A 2008, 'Pressing the flesh: a tension in the study of embodied, embedded mind', *Philosophy and Phenomenological Research*, vol. 76, no. 1, pp. 37–59.

Di Paoli, E 2009, 'Extended life', *Topoi*, vol. 28, pp. 9–21.

Gallagher, S & Crisafi, A 2009, 'Mental institutions', *Topoi*, vol. 28, pp. 45–51.

Gallotti, M & Huebner, B 2017, 'Collective intentionality and socially extended minds', *Philosophical Psychology*, vol. 30, no. 3, pp. 251–268.

Gergen, K 2010, 'The acculturated brain', *Theory and Psychology*, vol. 20, no. 6, pp. 785–816.

Ghosh, A 2016, *The great derangement: climate change and the unthinkable*, Harvard University Press, Cambridge, MA.

Hacking, I 2002, *Historical ontology*, Harvard University Press, Cambridge.

Hacking, I 2007, 'Kinds of people: moving targets', *Proceedings of the British Academy*, vol. 151, pp. 285–318.

Kono, T 2014, 'Extended mind and after: socially extended mind and actor-network', *Integrated Psychology and Behavioral Science*, vol. 48, no. 1, pp. 48–60.

Putnam, H 1981, *Reason, truth and history*, Cambridge University Press, Cambridge.

Rachlin, H 2014, *The escape of the mind*, Oxford University Press, Oxford.

Slaby, J & Gallagher, S 2014, 'Critical neuroscience and socially extended minds', *Theory, Culture & Society*, vol. 32, no. 1, pp. 31–59.

Taylor, C 1989, *Sources of the self: the making of modern identity*, Harvard University Press, Cambridge, MA.

Index

M. de Leeuw and S. van Wichelen (eds.), *Personhood in the Age of Biolegality*, Biolegalities, https://doi.org/10.1007/978-3-030-27848-9

R